YOUR CAPTAIN SPEAKING

Views from the flight deck

Captain Scott Birrell FRAeS

Wetsocks Books

ALSO FROM WETSOCKS BOOKS

Non-fiction

WALKING THE LINE
Two oldies (and one dodgy hip) tackle
the entire Western Front

Nick and Fiona Jenkins

LETTERS TO 281

Alan Jenkins

Fiction

DEATH & TAXES
(and other uncertainties)

Fergus Linnane

THE PALE PARABOLA
(and other stories)

Fergus Linnane

Dad's words

Think big and your deeds will grow,
Think small and you'll lag behind,
Think that you can and you will,
It's all in the state of mind,
Life's battles don't always go to the faster or stronger man,
But sooner or later the man who wins
is the fellow who thinks he can.
Walk tall, my son.

Papa

March 18th 1967

Published by Wetsocks Books,
Hebden Bridge, West Yorkshire

ISBN: 978-1-5272-7364-1

CONTENTS

When I turned for help and advice from my wonderful pals, none hesitated. Stewart Moffat, Ben Bamber, Keith Danvers, Colin Budenberg and Rob Moody all came up trumps and without their time and effort this book would not have seen the light of day.

I would also like to thank Leigh for the huge task of proofreading the book.

And to Ruth, who has always supported me throughout our marriage, enabling me to fulfil my wildest dreams in my flying career, who has brought the children up brilliantly, often on her own during the many absences when I was off flying, I want to say: 'I love you all around the world and back again!'

Scott Birrell

Bold? I'd rather be old

The Boeing B767 is unbelievably powerful. It is a long-haul, two-engine aircraft so it has to be massively over-powered. Taking off in a heavy B767 is never dull and never routine. I would taxi on to the runway for take-off and say to the guys on the flight deck: 'Are we ready, gentlemen?' They'd say yes, and I would add, just as the power came up: 'Then let's go on holiday!'

The next 30 or 40 seconds are exciting and critical. On take-off, as you are accelerating, particularly if you've got full power on, there is the most tremendous and exhilarating howl from the engines as you charge down the runway. You might still be on the runway as the jet is travelling at the best part of 200 miles an hour. You can't help but feel this buzz.

Strangely, the sensation of speed as we accelerated down the runway was not what I'd be thinking about as we approached our decision speed. All I was thinking was: 'Do I go? Do I stop? Do I go? Do I stop?' Past 80 knots, we would only stop for an engine failure, a fire, a wind shear warning or if I believed the aircraft couldn't fly. My eyes would be flicking between looking down the runway and inside at the instruments. As we reached V1, our decision speed, we needed to go. We had to fly, whatever happened, so I would be thinking: 'Go, go, go!' By the time the wheels and the flaps were up, I was pretty much disconnected from speed. I might be flying at four miles a minute or six, seven or eight. At that stage, it is just maths. If there is a thrill, it usually comes in relation to altitude. It certainly happened when I got up to 43,000 feet, which is as high as I've ever flown.

The sensation at that altitude is: 'Wow!' The sky above us is still not dark at that height - you would need to go up to 60,000 feet to experience that - but you can see the curvature of the earth and the fantastic cloud formations just by looking out of the cockpit window. The sensation of speed returns on the approach. As you get closer and closer to the ground, you seem to be getting faster. This is the phase where every pilot has to deal with the consequences of speed. Being aware of your proximity to the ground certainly holds your attention, even though the aircraft at that point will actually be flying at its slowest speed since take-off.

I love the B767, but even when I walk up to a Cherokee light aircraft now to go flying, I still get a tremendous buzz out of the fact that this sleek lump of shaped metal actually flies. I have never gone up to any aircraft that I'm going to fly without that excitement because I know that, little or big, this thing could actually kill me. I've always been excited at the prospect of flying an aircraft, but this excitement is always tempered by respect for what it is capable of. Even in a little Cherokee, if you get to 50 or 100 feet and the engine stops, you've got to hope you are going to be a bit lucky because at that point you may have nowhere to go. With luck, you won't be looking at a housing estate. I still feel that tremendous thrill every time I take off, even today. Back at school all those years ago, I had one plan and one plan only: I wanted to fly. In those early years, my ambitions were targeted in just one direction - I was going to be a flying instructor. With luck, effort and a bit of nous, I ended up flying the big jets. There is an aviation saying which goes: 'There are many old pilots and there are quite a number of bold pilots. But there are a dwindling number of old bold pilots.' I think about that sometimes, especially when I ponder the death of someone like my old mentor Jim Beaton, my first flying instructor. Jim's death was incredibly sad. He was well liked and hugely respected as a pilot, but he is still dead – dead in an aircraft, and a single-engine aircraft at that.

I got to the end of my professional flying career without, as one of my colleagues put it, 'screwing up', but I would be kidding myself if I ever felt that was the end of the story. I will always want to fly, but not big jets. I have gone back to my first love: light aircraft. I still get a thrill as I climb into a single-engined aircraft for a bit of local flying or a trip for lunch over to Le Touquet. But when I do, the lesson of Jim's death is never far from my mind. No matter how experienced, how competent or how skilful the pilot, one error of judgment is potentially just behind the next cloud or over the next tree line. You are being foolish if you treat aviation lightly. From a pure piloting point of view, there is a big difference between flying professionally as an airline pilot and controlling a little aircraft. In the jet, I always had a qualified crew around me and, as a result, it is a much safer place to be than in a single-pilot, single piston-engined aircraft, because the piston engine is considerably more unreliable than a jet engine. Jim was flying a piston-engine aircraft in North America in winter and heading out to cross the Atlantic with no radar and no de-icing. Flying like that, you are operating in an incredibly hostile environment and there is a good possibility that you are about to kill yourself. There's no doubt about it. The most experienced pilots of light aircraft in the world can make errors of judgment. The grave of famous pilot and record-breaking aviator Steve Fossett is proof, if that were needed. Anyone can make a mistake at any time, especially in a single-engine plane flying solo. There is no one there to say: 'Are you sure?'

So will I carry those rather dark reminders into my retirement from professional flying? Absolutely! I'm pretty competent, I know that. I can fly a little aircraft fairly well but I never let my 'seat of the pants' ability disguise the fact that this machine will kill me given a chance. If that happened, it would not necessarily be as a result of mechanical failure. Jim Beaton didn't die because of mechanical failure. He put the aircraft in conditions where it couldn't fly. I have talked to pilots within a training environment for many years about judgment. It was one of my mantras – judgment, judgment, judgment. Where catastrophic mechanical failure is not an issue, people in planes die usually as a result of poor judgment by pilots. Judgment will ensure that you always respect the elements, that you make every effort to understand the issues of the upcoming flight. My colleague who raised the issue of 'screwing up' was a senior captain and he made that comment at a drinks do to celebrate his retirement. I have thought about it a great deal ever since. Today I feel very much the same as that retiring captain, and that is a good thing to have in the back of my mind the whole time, constantly reminding me that I operate in a hostile environment.

I've been incredibly fortunate to sit at the back of the simulator or the back of the flight deck and watch excellent pilots - and not so excellent pilots - deal with many difficult and demanding situations and consider situational awareness, workload and decision-making. I have developed good judgment skills and these I have tried to pass on when training and checking other pilots. I am maybe not as bold as some of my colleagues. I am not saying that they are wrong, it's just how I perceive a difference in our judgment. If I diverted due to strong crosswinds and a colleague didn't, that wouldn't surprise me. Once again, it would just remind me that these aircraft will kill us all if the pilot makes an error of judgment. The retiring captain in the bar had spent his whole career flying. He was an ex-military pilot who joined Britannia and had a second career as an airline pilot, and as a captain he could walk away from the airline with his head held high. That's a very impressive record for anyone who has been dealing with the safety and comfort of 'the public' for many years in challenging environments. I reckon I managed to do the same… against the expectations of some of the people who formed judgments of me in my early years.

Don’t be stupid, boy!

He had always been a miserable bastard. But as an example of the benefits of a private boarding school education, I was not exactly the headmaster’s model pupil. In those days, there was no word in the Scottish dictionary – or any other dictionary, for that matter – for the ailment which was to become known as dyslexia.

So, to my grisly headmaster, this bespectacled, overweight boy would pass from the gates of Keil School, Dunbartonshire, with nothing much to recommend him – academically backward and lacking any kind of prowess at sports or on the rugby field. Nevertheless, before bolting for the school gates marked Exit – never to return – I felt it proper to present to this awful man a small token, perhaps imagining that even at this late hour, he might at least remember something good about Scott Birrell.

If I had expected gratitude or any semblance of warmth from this Scottish relic, my optimism was quickly dashed. After I timidly handed the headmaster a bottle of whisky as I said my goodbyes, it seemed for a moment that he might at last be offering me something approaching civility, when he enquired: ‘So what are you going to be, Birrell?’ With renewed affection for the old boy rising within, I replied proudly: ‘I’m going to be a pilot, sir.’

Reality struck back with a vengeance when he snarled: ‘Don’t be stupid, boy!’ I have never forgotten, and I have never forgiven, that comment. Those words remain etched across my heart to this day.

It wasn’t, of course, my earliest memory. The first thing I can remember is being in one of those old-style prams with a cover pulled over the top and looking through the little tassels. I was born on March 18th 1955 in Glasgow and we lived in a bungalow in Giffnock, which was a nice suburb. I remember Mum telling me she was going to have another baby. My brother is almost three years younger than me so I must have been about two and a half. I was the only child at that stage.

My father was a chemist and optician and ran his own shop and business in Govan, the shipbuilding part of Glasgow. My mother was a South African who had run a dance school in Johannesburg. She began dancing classical ballet but went on to do all kinds of shows in the UK. She and film star Julie Andrews, who made The Sound of Music and Mary Poppins, were pals and they were in shows together in England in those early years. So Mum came to mix with a lot of interesting showbusiness people - she even met American film star Mickey Rooney.

It was while she was travelling with one of her shows that Dad first saw her in Glasgow, where she was performing in a musical. He became so enamoured that he pursued her around the country, armed with large bunches of roses. Eventually, and perhaps inevitably, he won her over. They married on March 28th 1953 at the Church of Jesus Christ of Latter Day Saints, in Claremont Terrace, Glasgow. After marrying Dad, Mum gave up her showbiz life. My father had been married before, although he never told me about it. I only found that out just before he died, from his sister.

It was quite a strange family, really, not that I thought about it at the time. My mother had a sister Lillian, who died of tuberculosis in South Africa, and my father had a brother and a sister. But those two never married and they ended up living together. They were my uncle and my auntie. I didn't have an uncle 'this' or an uncle 'that', I just had one uncle and no cousins. They lived together in Glasgow and they both ended up working for my Dad. My Dad had another brother named Scott, who went to America. I can only recall ever seeing one picture of him which was taken on board some kind of cabin cruiser. I have no knowledge of what happened to him.

My father, my aunt and my uncle all seemed to have tremendous regard for their mother, who had been given a bit of a tough time by her husband. They rented a big tenement building in Glasgow, and I guess were regarded as working class. They certainly weren't living in poverty because my grandfather was an engineer.

My mother's mother was from an Irish line and her father was from a Scottish line, which extended a few generations back. Her father had been very successful and had managed to make and lose a fair bit of money a few times in his life, but the family story was that he gambled it all away in one night. Cyril also bred Pomeranian dogs and had significant mining interests at one point, according to his brother Hal, whom I eventually met a few times.

During the Second World War, before Mum moved to the UK, she enjoyed a good life in South Africa. She was brought up by a nanny, and when she came to this country, she couldn't even boil an egg. But maybe due to the ups and downs of their financial situation, or possibly because of my grandfather's eventual infidelity, there were problems in the family. So when my Mum came to this country, after winning an important dancing award, her mother decided to come too.

We lived in a good-sized bungalow with Mum and Dad, my grandmother and, on top of that, my aunt and uncle were around the house a lot of the time too. My aunt and uncle lived in the Charing Cross area of Glasgow where they had a big flat in one of the old tenement buildings. They would come around to our house and stay for long weekends, and at Christmas they would visit and not go home until March, which eventually drove my Mum quite potty.

In my earliest years, my cot was placed in the room in which my Dad had his own little bar. Eventually, I ended up in a bedroom and, by the time my younger brother Nigel came along, he and I had a double bed to share. However, they decided that our double bed was to be beside the single bed where my grandmother slept - and that's where it remained until I left to go to boarding school, aged fourteen.

When I was eleven, Mum decided that she wanted to adopt a girl, and a little baby called Gail arrived in the house. In those days, the law was that the biological mother was allowed to take the child back at any time during the first six months, if she felt that was what she wanted. One day, coming home from school for lunch, I found Mum in floods of tears. They were taking the baby back after five months. What a terrible, terrible thing.

We had a cleaner who used to come to the house every week to help my Mum and it turned out that her niece was pregnant and was due to go down to London to give birth. Dad said to me: 'I am not having your mother going through this again. I will not accept that.' My next recollection is being in the back of a car with my Dad and sitting opposite were people I'd never seen before. Dad had an attaché case on his lap, and when he opened it up, it was full of money. Dad shut it, passed it over to this person, and said: 'If you come for the child, I'll kill you!' This was in Glasgow, and the next thing I knew, Mum was going off in a BEA Trident to collect the baby in London. When she came back, I had my wee sister Traci, just five days old. Nigel and I always said that Traci came by a Trident, not a stork.

Looking back, I was at that time, I guess, a little bit wary of my Dad. He was a powerful man, quite a big man, and in his shop he was regarded as a sort of god-like figure. Sometimes I would be in the car with him and we would be stopped by the police. It was always quickly sorted out and we would be off again. When I asked him what it was all about, he said: 'Oh, I keep a case of whisky in the boot and they just help themselves.' On another occasion, he wound down the window and said to the policeman: 'Do you know who I am? I am Jim Birrell.' And off we went.

My father was an Edwardian sort of chap. I remember him telling me: 'Your role in life, Scott, is to get a job and provide for the family.' I would have been sixteen when he said that. Interestingly, it was not: 'Your job is to be a good husband, to be a good father.' He worked practically every day and no doubt felt he was fulfilling his concept of a husband and a father.

Every so often, as a family, we would go away down to Ayr or Largs on the coast – Mum, Dad, Auntie, Uncle, Nanny, Nigel and me. We would stay in a nice hotel, where Dad would go into the bar and drink. Mum would only have a soft drink, since we were brought up as Mormons and she didn't drink alcohol at all, but she would dress quite glamorously – she would always stand out in a crowd of Glaswegian people.

When it came to dining out, I remember Dad telling me at quite a young age: 'There is no point in tipping at the end of the meal. You might as well tip at the beginning because then you are going to get good service.' He had a big moustache and would walk in wearing an immaculate white dinner jacket. No one wore white dinner jackets in the 1950s in Glasgow. Dad might have been considered a bit flashy in those days.

We went on holiday to Viareggio in Italy. Much of Europe was still feeling the effects of the war years and Dad said that people we met driving through Europe in the Jaguar had never seen such a grand car. It wasn't until much later in life that I used to wonder how Dad could have afforded it all. The Italian trips went on until I was about eight and then Mum and Dad stopped coming altogether. But for us boys nothing much changed because every year my auntie and uncle took us instead. And we went and stayed at the same hotel and did the same things every summer for quite a few years.

Aged five, I was sent to Giffnock Primary School, where I would stay until aged 12, when I went to secondary school. I hated school in the early years. I was bullied and I used to run away and go back home. I had some good pals but this still didn't prevent school being pretty unpleasant – until I found Billy McBean, who was a big boy for his age. I quickly discovered that by giving him two shillings on a regular basis, he would agree to become my protector. From that moment on, all bullying stopped. Sadly, if I thought I was miserable at my first school, my second school turned out to be much worse. There used to be knife fights in the playground. There was always the Catholic-Protestant thing going on in Glasgow and it was endemic in the secondary school. Fortunately, I quickly worked out that when people said to me 'What are you?', I would say: 'I'm a Mormon.'

The secondary school was called Wood Farm and was a big comprehensive. Needless to say, I didn't like it at all and I struggled. I don't think I really understood what school was for. I was well behaved, by and large, I didn't cause any trouble, but the playground was nothing less than a battlefield. So one day I went home and said: 'I want to go to boarding school.'

So my fourteenth year was quite a year: as well as Traci being born, I started at boarding school. More important than that, it was the year in which I completed my first flight.

Nanny who lived with us, had been converted to the Mormon faith in South Africa. The Mormons were very passionate about delivering their message to people on the doorstep. My mother and my grandmother were both converted at the same time and were baptised on November 17th 1931 at Mowbray, in Cape Province, South Africa. My grandmother gradually lost interest but my mother remained extremely strong in the church, as are my sister and sister-in-law. So I went to church.

By the time I got to fourteen, I was expected to go to church when I came home from boarding school but I used to get fed-up with being asked every time I went to church: 'Oh, would you do this, or would you do a two-and-a-half-minute talk on something?' It started to put me off the whole thing. One day Mum came into my bedroom and said: 'It's time for church.' And I said: 'I don't want to go.' I didn't stop believing in it, I just drifted away. When asked now, I still say 'I am a Mormon' – even when I'm holding a glass of wine.

My boarding school, Keil, was on a fifty-acre site on the banks of the River Clyde, originally the property of the shipbuilding family, Dennys of Dumbarton. It was an independent secondary boarding school with 200 pupils and a few day boys. In 1978, it became co-educational, and it closed in July 2000. The site has since been developed for housing.

Arriving at the age of fourteen, I was starting in the third year. Most of the boys had arrived aged twelve. I arrived at Keil with Mum on my first day, a beautiful late summer's day. We drove through the grounds and up to the large house, Isley Kerr, which was to be my house during my time there. There were lots of cars arriving and boys pulling large trunks out of the boots. Taxis brought boys from the airport back to school after visiting their parents in faraway lands. A lot of the boys' fathers worked for the Foreign Office. Everyone finally went and I was left alone with this bunch of boys I didn't know. I felt homesick. I phoned Dad one evening, saying: 'I don't like it here. I want to come home.' Dad said: 'Look, son, you've just been there a short time. Why don't you give it a try till the end of the term? But if you're really, really unhappy, pack your bags and I'll come and get you, or you can go and catch a train.' That conversation made a huge difference and I felt much happier.

I very quickly realised that I had landed in heaven. Once I got over being away from home, I found out that three times a day, I was going to enjoy what could only be called 'real food' after my mother's poor cooking. As I grew up, I had started putting on weight. By the time I went to Keil at fourteen, I was up to sixteen stone. At one point, it was noted that I was a bit overweight, so I was put on a diet. The cook would bring me something less fattening. This was marvellous, because I'd eat that – and then I'd eat my normal dinner.

The teachers were an interesting bunch. One day, a new master arrived. He came into the classroom and he didn't say anything. He just went up to this tall teacher's desk, opened it and took out a nail, about two inches long. Then he got out his strap and hammered the nail into the desk. By the time he had achieved that, the class was sitting in complete silence, mouths open, all thinking the same thing: 'If he can do that to a nail, what can he do to our hands?' The standard punishment was to get belted with our hands held out in front of us, two hands crossed together so the strap didn't hit your inside wrist, causing it to swell up. Despite this threat, he turned out to be a brilliant teacher. He just had the panache to think of a dramatic demonstration to keep us in order.

I was no good at sport. We all had to do PE but there wasn't much in the way of facilities. We'd go for runs or practise rugby, although I never did know the rules. I was quite big for my age - weight-wise as well - but I still found rugby a bit too much rough and tumble. I'm not really into all of that. To this day, I don't do any sport. I can remember the Latin master, one frosty day when the ground was rock hard with frost, saying to me: 'Birrell, dive at the ball.' And I was looking at him thinking: 'Are you on drugs? Do you think I'm diving at the ball on concrete?' Summer holidays were long. We had nine weeks off and, of course, I enjoyed that because it meant I could go off and fly. How did I do at school overall? I tried. I was not bad at history, physics was interesting, and chemistry was okay. Also I was not bad at art. I was dreadful at maths but fortunately, to fly a jet much later in life, all I would need was my three and eight-times tables: the three for altitude for descent planning, and the eight for speed, for example, eight miles per minute.

Mum had a Morris 1100 four-door in which she taught me to drive when I was fifteen. I can still remember driving to Oban in that car. It didn't seem to bother her in the slightest that it was illegal. I got my first scooter at the age of sixteen. It was a brand new Lambretta, but it wasn't awfully robust and it kept breaking down. Even today I have a pin in my left arm and my arm won't straighten. My pal was riding the scooter and I was on the back when he took a corner too fast and we hit a wall. I ended up in hospital and had to spend the night there. I was sixteen. I remember them saying at one point that the pin would come out, but it never has. I could drive a car legally by seventeen, so as soon as my birthday came around, I started driving lessons and passed my test by the end of the month. Dad had lent me Mum's car to learn in and then, when I passed my test, they bought me a Singer Chamois, a version of the Hillman Imp. Not many boys could drive in the sixth form and none of them had their own car. That car used to drink more oil than fuel and proved to be pretty unreliable. Dad had bought it from a garage in Govan for £175 and I was always very grateful because, despite its unreliability, to me it represented freedom.

Around that time, Mum was saying how disappointed she was that I'd gone to boarding school, on which they had spent a lot of money, and now I had begun misbehaving. By that, I guess she meant I was no longer interested in the church, I was drinking, and I had not done terribly well academically. She felt it had all been a waste of money. I remember saying to her: 'It wasn't.' Because I knew it wasn't. Keil had actually been the making of me. This school had taken me from someone who was a bit lost in the world, someone with nowhere to go, someone who was a bit fragile, to someone who felt he was on a very progressive journey. My entire last year was really just about waiting to finish, and in the end I didn't wait for the last day of school before I left. I knew I didn't need any school qualifications to become a flying instructor.

I'd become aware that some of my chums were starting serious relationships with girls and I wasn't. Even though I had had my first kiss with a girl from a school in Helensburgh, I had already begun to find out that flying and the world of aviation were much, much simpler activities than girls and relationships. I guess it was a bit unusual that by the time I went to Keil, I had actually started flying lessons. I'd had two lessons before I started at school.

As I walked out of school for the very last time, I thought: 'I'll never, ever come back!' And with a skip in my step, I left Keil behind forever.

Love at first flight

One year, during the school summer holidays, we took one of the old steam trains from Glasgow to London, and then out to Heathrow where we flew to Nice with BEA in a Comet. This was the first time I'd flown. In those days, Comets were used for both long and short-haul flights. Nice was fascinating because there, at the airport, I could watch the Air France Caravelles, with their strange little triangular windows. We did that trip year after year and it was during one of those early trips that I noticed how the runway at Nice ran along the seashore. I'd look out of the Comet window when we landed, looking out to sea.

I can still remember the excitement of those early flying days. Dad would drop us off at the airport, if he could not leave his business, and Nigel, Mum and I would head off on another adventure. Once we flew in the Comet from Heathrow when it had an engine failure while we were in the air and we had to turn back. I had fallen asleep, and I woke up for the landing with Mum explaining: 'We're back in London.'

On one departure in a British Eagle Bristol Britannia, it was raining heavily and the pilot did a 'wet start', as I now know it to be. I saw this big flame shooting out of the back of the engine, and I was off running down the aisle, shouting something to all the passengers in a panic. The air hostess had to sweep me up and put me back in my seat.

The first time I was allowed to go on to the flight deck was aboard a BEA Vanguard, on our way to London. I was about eight, and I was allowed to go in and see the pilots. I thought: 'Wow! I want to be a pilot!' It wasn't just the aeroplane cockpit that intrigued me; the whole flying environment was very special and exciting. I can still smell the old Viscounts and Vanguards. These days, if we go to Duxford Air Museum and there's an old Bristol Britannia, a four-engine turboprop, on display, it will still have exactly the same smell I remembered. Modern jet aircraft somehow have a different kind of smell. It is different but nonetheless exciting. I used to recognise, and be excited by, that smell of burned kerosene.

My auntie used to take Nigel and me on trips to Renfrew Airport. She didn't drive so she would take us by bus and then, when it closed, to the new airport, Abbotsinch, an old naval station which is now the site of Glasgow International Airport. We'd have lunch there and we would stand for hours and look at the aeroplanes. We'd go up to the desks in the terminal building and ask for little maps and airline flags which were on the counters – BEA, BOAC, Air France, and Swissair. We kept them all in our bedroom.

I had started to build models at about ten, but it was military stuff and I was not so interested in that. To me, it was the airliners I was mad about. I knew the shapes, colours and the names of all of them.

When I reached fourteen, during the summer holiday before starting at Keil, I saw an advert in a newspaper saying 'Pleasure Flights from Glasgow Airport'. It was with Loganair, a small Scottish airline I eventually flew for many years later. The trouble was, you needed £8 for one flight. So myself and a bunch of pals – about eight of us – decided to put in £1 each and cycle out to the airport. We got there, I said 'Wait outside' and I went in to organise whatever I had to do to get us on the aeroplane. I went through the wrong door. It was the Loganair building, but I had walked into the flying club entrance without knowing it. Bold as brass, I went in and said to the first person I saw: 'I've come for a pleasure flight.' The chap said: 'We don't do that. We teach people to fly.' This struck me like a bolt of lightning. I remember thinking: 'Teach people to fly!' As a flying instructor myself, many years later, many kids would walk in, just as I did – and they were a bloody nuisance. But this particular guy was Jim Beaton, who turned out to be the club's CFI, chief flying instructor. He, more than any other person before or since, was to transform my life. He said: 'Yes, we teach people to fly here.' I must have asked some sensible questions after that because he and I chatted for ages. My pals, waiting outside, never did go on a pleasure flight. I asked Jim: 'How much is it?' He said it would be £6 or £7 an hour to learn to fly. He showed me around the hangar and the different aircraft – 'this one is £6 or £7 an hour, this one is £8 an hour'. The first was a Piper Cherokee 140, a single-engined, four-seater aeroplane, and beside it stood a Bolkow Junior, a two-seater.

It was just amazing. Standing next to an aeroplane, I was just blown away. Did I sit in it? Possibly. Jim spent ages with me and explained all the things I would need to learn and what I needed to do. He said you've got to be eighteen to join, possibly because the Glasgow Flying Club had a licensed bar. Then he asked me: 'Are you really, really keen?' I said: 'Yes, yes.' So that night, very kindly, he phoned my father.

After the conversation, Mum and Dad said: 'Fine, we'll help you a little bit, but you will have to work for the rest.' They probably thought: 'It's not a bad thing for the boy to be doing.' Mum, Dad and I met Jim in the Tinto Firs Hotel in Glasgow. Jim was quite a mild man, not a drinker. He explained that I'd get a private pilot's licence, but I couldn't go solo until I was seventeen. I was only just fourteen. The meeting with Jim went well, except that when we came back home, Dad said: 'He's got short arms and deep pockets' - Jim hadn't bought many drinks. I was a bit horrified that Dad was being nasty to my new favourite person in the whole world. What mattered was that Dad had given it his blessing. Because of the bar, I couldn't join the Glasgow Flying Club, but Jim started the Beaton School of Flying – and I was the only member.

I remember very clearly my very first flight. I was in the left-hand seat – the pilot seat – which is where you sit when you are learning, and Jim was in the right-hand seat. We were in the Cherokee, G-AWEV, and all I could see were dials everywhere. Glasgow is a big, busy airport, so the environment is very controlled and regulated, with lots of talking to air traffic control. Jim was calling for start-up and there was a lot of chatter going on. Of course, I couldn't understand a word of it. I was sitting on a cushion and realised that to steer the aeroplane on the ground, you had to do it with your feet. I followed through on the take-off with my feet and my hands on the controls. It was joyous and exciting - and just a little bit scary.

We flew off to Loch Lomond, and during that flight Jim let me fly the aeroplane in a straight line, straight and level, then a bit of turning. I built up to this nice and slowly. He showed what would happen if I lifted or lowered the nose. And then I was sick! No doubt it was a bit choppy and the next thing I knew, I was throwing up. I threw up every time I went flying. Jim had to clean up after his star pupil had pedalled off home on his bike. Eventually, he told me to clean up the aeroplane myself!

Of course, things got a whole lot more exciting after I met Jim Beaton. From that moment, my whole life changed dramatically and I wanted to spend as much time at the airport as I could. That long holiday of 1969 was a turning point in my life. I was starting my new school in September and I was quite happy that whatever happened from now on, my life was not going to be just more of the same. At the flying club, I had become a bit like the club mascot. My exploits were even written about in a Scottish newspaper. There was an elderly gentleman who had been involved in motor racing and flown Tiger Moths back in the 1930s. The article was about the youngest pilot – me – and the oldest pilot in Scotland.

There were two flying clubs at Glasgow airport: the GFC and the West of Scotland Flying Club. The Glasgow Flying Club flew Piper Cherokees and the other club had Cessnas. The Cherokee was a low-winged aircraft and the Cessna a high wing. At that time, 1968/69, the Cherokee was quite a new aeroplane, not that I knew it at the time. They also used to have a Piper Tri-Pacer, which had a tricycle undercarriage and a high wing, but the Cherokee was the next generation. So there were a couple of Cherokees, a Bolkow Junior, which was a bit tatty, and other aeroplanes which came and went over time. I would fly an hour every two weeks to begin with, but I was taken up on lots of other people's flights. If someone went flying, they would say I could sit in the back. We'd fly down to Prestwick and would sit on the grass and sunbathe and watch everything going on around us – you can't do that now.

I spent every moment I could at the club, just listening to the pilots talking about aviation. When Flight International came out, a weekly aviation magazine, at the back there were advertisements for different flying jobs. We would wander over to the terminal to get a copy, then sit in the coffee bar and I would listen to the other pilots talking about all sorts of things: everything from Robbie Hunter's favourite shape of women's breasts – which I remember him drawing on a board – to the best short field take-off technique. Robbie went through pilot training with the RAF and was offered a navigator's course, but turned it down. That was quite common in those days. I guess it depended on the RAF's requirement at the time, but he certainly did fly Jet Provosts at one stage during his pilot training, so he was a sort of god, as far as I was concerned. He had his CPL/IR - his commercial pilot's licence and instrument rating.

Ray Connolly was a private pilot and also an air traffic control assistant. All these men were extremely interested in flying, but from different backgrounds, and they would invite me along because they knew I was passionate about flying.

Jim and I flew across to Edinburgh once, and after landing someone told Jim off because we had not joined the circuit properly (a circuit comprises of designated routes to be flown visually for take-off and landing). We had apparently joined on the wrong side, and Jim said: 'I was asleep, it wasn't me.' He pointed to this fifteen-year-old boy – me! Scotland's airspace was mostly unregulated at that time. There was almost no controlled airspace, other than upper airways which we were not allowed into, and couldn't reach anyway. Even Aberdeen and Inverness were in uncontrolled airspace. Edinburgh, Glasgow and Prestwick were, but that was about it. And because it is such an open country, and because there's not that much flying going on, it is unlikely that people on the ground would get annoyed with light aircraft overhead. I became a bit of a hooligan in later years with an aeroplane. In those days, I was still just a young person learning to fly, and we would go to Mull, about an hour and a bit away. It was a fabulous flight because you'd fly down the Clyde, over the Crinan Canal, pass the Corryvreckan Whirlpool and then up the coast, towards Oban, and across and down the Sound of Mull. Mull was a lovely island with a wonderful airfield called Glenforsa, a little grass airstrip between the hotel of the same name and the Sound of Mull. The airstrip was right on the shore, running east-west.

The Glenforsa Hotel was about halfway along the strip, and you could taxi up a wee hill and park by it. Behind the hotel was a glen, going up into the hills, and that caused a lot of problems because if the wind was going up or down the glen, it came right through the middle of the strip. So quite often the windsocks at each end of the runway would be pointing in opposite directions – not what a pilot wants to see.

It meant it was quite an awkward place to land. The strip sloped down to the sea, so the aircraft, if you let it do what it wanted, would gently turn down towards the sea. The grass, when wet, didn't give much traction, and awkward winds could give you a tailwind on landing and a headwind up the other end. Occasionally, there would be a very awkward crosswind to contend with. So, unsurprisingly, there used to be quite a few prangs on Mull. The club introduced a rule that you had to have a checkout by an instructor.

People would fly up from England because they had booked into the hotel, which was in a fantastic location. It was a modern Scandinavian log building owned by the Howitt family. Old man Howitt was an engineer who, it was said, had played a leading part in building the early stages of the M1. The Howitts had three offspring who were all older than me, two boys and a girl.

David Howitt was the youngest son. The strip was operated by various councils and, during my time there, was licensed by the Civil Aviation Authority (CAA). The Howitts had the original hotel, which burnt down some time after the airfield was built by the Royal Engineers in 1965. By the time I was flying up there, the new hotel had been built. David and his brother Tim manned the old fire engine, parked by the strip. The fire engine was a really old banger which seldom moved. It certainly would have been of no use to you if you were burning. Mull was possibly the first grass landing I'd ever experienced, and later on I managed to crash my own little aeroplane quite spectacularly there. One of the first things I learned there was a short take-off technique, using a bit more flap and getting the nose wheel off the grass and rotating a bit early. It wasn't a very long strip, and the environment around the strip was pretty treacherous.

I quickly discovered that the whole family enjoyed a drink and they were all real characters, which I thought was marvellous. I took to David in a big way. He would work a little hand-held radio for air traffic: you'd call up and he'd tell you what the wind was doing. He would say 'I think it's okay to land' or 'it's looking a bit this or that'. There was a Loganair service into Mull from Glasgow with a Trislander, and the air ambulance, an Islander, would also occasionally fly in. Small wonder that I'd fly there as much as I could. I'd go along if anyone was going up, or if there was a 'fly out', as we called it, when maybe four or five aircraft would fly to Mull for the day.

You couldn't put four people in a Cherokee into Mull because the runway was short and the grass further degraded take-off and landing performance so there would be just two or three of us. When other people were paying for the aeroplane, they wouldn't put me in the right-hand seat, I'd be in the back, but if Jim had to move an aeroplane somewhere, I'd quite often get to fly it.

Robbie Hunter, who was a commercial pilot, flew a Piper Aztec, which was a two-engined six-seater. They were very popular as air taxis. As I got a bit older, Robbie would take me with him in the Aztec if he was delivering a piece of freight somewhere. On one trip with Robbie to Paris Le Bourget, we took a hospital scanner in the back.

I was slowly becoming recognised at the airport. I knew a few people at Loganair, Geoff Rosenbloom and his small team at Air Charter Scotland, and most of the pilots at the Glasgow Flying Club. By now, I was starting to plan how to fulfil my ambition to be a pilot, and that meant becoming a flying instructor or, to start with, an assistant flying instructor because I didn't need school qualifications. At that point in my life, the whole idea of school and education became, to me, completely irrelevant because I didn't need it to move forward in my chosen career. You can't do it that way now but in those days you could become an airline pilot without any formal qualifications or professional courses. You could home-study the whole lot. We were known as 'self improvers', and not highly regarded by many. Usually, when you learn to fly, you start with the effect of controls to see what the primary and secondary effect of each control is. Then you practise straight and level flight, and introduce turning. After that, you move on to climbing and descending, then climbing and descending turns. You progress on to stalling. In those days we used to spin the Cherokees - no longer in the syllabus - although they didn't spin very well… just enough to scare the shit out of me! By the time I could do all of that, it was time to start circuits and to practise take-off and landings.

Eventually I was doing the take-off and landing with Jim following me through on the controls. That would have taken four or five hours up to that point, but there was never any rush because I knew I couldn't go solo until I was seventeen. In the meantime, sitting in the back of an aeroplane while someone else was flying the circuit was amazingly useful, because I'd be listening to all the tips and the patter from the instructor. He would ask the student 'How did that go?' or 'Why did you think that happened?'. We didn't have headsets in the aeroplane, just a microphone and speaker. Headsets were unusual and very expensive. I could hear all the chatter because I was right behind them, and I would be learning as I went around.

I learned that Runway 24 at Glasgow meant the runway was 240 degrees from magnetic north, but when we were flying downwind, it took me a long time to realise that the 060 heading was the reciprocal of runway 24. I was never the brightest button.

Once I could take off and land, I'd practise them by doing 'touch and goes' – landing, then immediately taking off again without stopping. After learning all that, it would normally be time to start going solo but, of course, because of my age, I couldn't. Nevertheless, Jim continued my progression by introducing me to cross-country navigation. We would sometimes fly off somewhere like Strathallan or Edinburgh for coffee. Jim was always very generous when it came to arranging for me to fly free of charge. If an aircraft had to be moved somewhere for engineering work, he would let me fly it. That way, when we flew a two-hour sector between, say, Glasgow, Prestwick and Edinburgh, a lot of the cost would be covered by Jim or the club.

I was shown a piece of equipment called a CRP-5 computer, which looked like a circular slide rule wheel and was used to calculate track, heading and ground speed. You would be given the 2,000 foot wind from the Met office, and then you'd plot the route on your quarter mil or half mil map, which was plasticised, so you could draw on it with your chinagraph pencil. You would establish the track using a protractor. Then, using the CRP-5 wind grid, you'd work out your headings and ground speeds for each sector. With each ground speed, you could then calculate the ETA for each way point, and your drift angle would produce the heading to achieve the track. Once airborne, you'd use the 1 in 60 rule (a 1 degree error after 60 nautical miles equals 1 nautical mile off-track) to confirm your calculated heading was correct. Jim taught me all of this, but I was more interested in just getting in the aeroplane to go flying.

I didn't always fly with Jim. There was another nice chap, a first officer with British Caledonian, who would come in and do some instructing. By the time I got to fifteen or sixteen, I was able to do so much myself, but while I was quite proficient, there was not much self-discipline and, I guess, I was suffering from complete overconfidence. To me, aviation became a place to go and play.

There were some fantastic characters all around me. There was a chap who bought an old BEA de Havilland Heron, a four-engined little airliner with a fixed undercarriage. BEA used to fly them around the Highlands and Islands and used the Heron as an air ambulance before Loganair took over the service. Then there was Davie "Jock" Adams, a BAC 1-11 captain with British Caledonian, and renowned for being a bit of a hooligan and a terrible man with the women. Later on, he and I were to get into various scrapes. I'd see him when he had flown in for the night, up from Gatwick, and he'd come to the club. He and others like him were the reason the club was such an exciting place.

I was seventeen and just after my birthday I went on my first solo flight. Normally, this would be achieved after eight or nine flying hours, maybe ten, and some people would end up doing it after sixteen hours. It took me a bit longer to do it because the rules stated you couldn't go solo before seventeen. In addition, a student would need more experience and a higher level of competence flying at a busy international airport, like Glasgow, than at a small grass strip. There was loads of commercial jet traffic flying in and out, so you had to have good RT discipline and confidence. When they told you to 'hold', you had to be able to do just that: hold downwind by orbiting left or right as required.

Your first solo is always one take-off and landing with your instructor witnessing from the control tower. I didn't know when my first solo flight was coming, but one day we were doing touch and goes and Jim said: 'We'll do it tomorrow.' When we taxied in, he said: 'Just keep the engine running.' Then he called up on the radio and said: 'I'm going to send someone first solo.' They acknowledged that and approved the request, and he climbed out on to the wing. He kept the door open, the engine was ticking over, and he dipped into his pocket and pulled out a box of matches. He tossed them on to the seat beside me. 'If you bend it, burn it!' he yelled as he shut the door.

I took off and remember going downwind for the first time with no one beside me. I was remarkably calm. Jim had somehow taken away all the hype from this landmark event. I completed my first short solo flight, did my landing and came in.

After more solo circuits and dual cross-country flights, it was time for my first solo cross-country. I was to fly a triangular cross-country, Jim's instructions being to fly to Cumbernauld, on to Loch Lomond, and then back to Glasgow. All very straightforward. As I approached Cumbernauld, I began to lose interest in the navigation exercise and was drawn towards the loch. Most people on their first solo cross-country would have quite rightly been saying to themselves: 'I've got to get this so right.' But I turned towards Loch Lomond and began flying along the Campsie Hills to the loch, whereupon with a rush of excitement I pushed the nose down and dived to a level about fifty feet above the water. With a huge grin on my face I was whizzing along at low level over the loch, over and around the islands, completely illegally. A hooligan was born!

This revelry lasted only a few minutes. Out of the blue, as I was banking over one island, another light aircraft, also banking hard, skimmed past me from the opposite direction. The instant it sped past my wing, the radio crackled on and a voice barked: 'You're grounded!' It was Jim Beaton – he was after me. Shocked and not a little annoyed, I pulled up into a more regulation flight path and pointed it in the direction of Glasgow Airport, all the time trying to think of a suitable excuse. It would be no use, of course. Grounded, he had said, and grounded I would be.

Back at the club, the bollocking began. He had gone up to the control tower to the radar suite as soon as I took off, and from there he saw me disappear off the radar when I was over the loch. I was flying one of Jim's own Cherokees and he had immediately sprinted from the control tower, hopped into another of his Cherokees, and sped off in pursuit. He knew exactly where to find me, he said. He knew I would do it. These days, I have all my log books and records, so it's fun to look back and see exactly what happened. I took my bollocking and resolved to try to behave myself in future. Or, perhaps more accurately, I resolved to not get caught misbehaving in future. My first cross-country solo had been not long after my first solo flight, simply because at that point there was nothing for me to learn as part of the curriculum. After all, I had been learning since I was fourteen and I had flown all over Scotland, with quite a few visits south of the border. All the things you needed to learn, I had done time and time again. If, for example, Jim said 'okay, so now we are going to do steep turns', I had been practising those for three years. So I could handle the aeroplane fine. You are meant to do a series of written exams to get your PPL (private pilot's licence), and I didn't really do that. I might have done a couple of exams and I passed everything. The navigation exam was supposed to be the toughest one. I didn't get too much formal instruction in the subject, but Jim got me through. And then it was time for me to go to Edinburgh to do my PPL flying test, because although Jim could instruct me, he couldn't test me. I had to be tested by a CAA-authorised examiner.

I flew there solo with the aeroplane and met this old gentleman, Mr Drummond. He looked at my log book and found that instead of the 40 flying hours he would have expected to see, I had maybe 80 or 90 hours. Every time I had flown with Jim, the hours had been logged. The examiner said: 'How many hours did it take to go solo? It's taken a lot.' I said: 'I started when I was fourteen.' He replied: 'Oh, that's all right then.' I did my test, which lasted 45 minutes, and there I was, a qualified PPL. I flew back to Glasgow with the greatest feeling of joy, satisfaction and achievement I had ever felt.

My new PPL allowed me to fly by day in good flying conditions, and take passengers - not for money, but I could now take my friends up. I could take three passengers in the Cherokee. I took Dad flying once and he was quite impressed. At home, passing my PPL wasn't such a huge drama because I had been flying for so long. Possibly Mum and Dad couldn't see it yet as part of a future career - even then, most airline pilots had degrees - but they were always encouraging, right from the first time I said: 'I'm going to become a pilot, a flying instructor.' I never wanted to be a commercial airline pilot because I believed in my heart of hearts that I couldn't achieve that. To my mind, you needed to be clever and I knew I wasn't. I had been told it often enough. To become an airline pilot in those days, either you were sponsored by BEA through their course at Hamble, you could pay for a course which cost a fortune at Perth or Oxford, or you could train in the military.

Even when you had completed training with the military, you still wouldn't be qualified as a commercial pilot. You still had to do all the CPL/IR exams. To me, looking at it realistically, that was just unattainable.

A word from the wise

Mum and Dad were disappointed with my academic results. This didn't matter to me, but it certainly mattered to them. I had found a way to end my schooldays with a clear plan for the future, and the real possibility that I would eventually create a successful flying career for myself.

I just knew without any great concern that I needed to do something to get things moving. By that time, the Singer had died a sad death. Luckily, Dad was persuaded to buy me a Mini. He gave me a fixed budget, which meant I could either have an 850cc saloon or a 1000cc Mini Van – so I went for the bigger-engined Mini Van. My pals helped me do it up a bit and this was a typical boy racer's car. Unfortunately, I crashed it. It was in the countryside at night and I was overtaking another car. There was an oncoming car in a hidden dip and we collided nearly head-on, with me on the wrong side of the road. The other driver was hospitalised. I had only a few minor cuts and bruises as I had a full harness fitted to the car's racing bucket seat.

I went off to Langside College to have another go at my maths and English qualifications. It was Mum's idea but I wasn't really interested. On the day of the exams, I walked up the steps to the college and looked up at the sky. It was clear blue and I thought: 'Let's go to the airport.' So I did. When I got home that night, Mum said: 'How did the exams go?' I said: 'Okay.' I never did tell her what really happened, and she never ever brought the subject up again.

I worked as a baggage handler going into my eighteenth year and I really enjoyed that experience, but my crash in the Mini made Dad declare: 'You're on your own now when it comes to cars.' I managed to buy an old Mark 1 Cortina with a knackered engine which could just about make it to the airport. One day, it was stolen from the airport staff car park, but unfortunately the police later found it. I was baggage handling for six months, working hard and not taking leave, but I was pretty tired and thought: 'I'd like to go on holiday.' One day, out of the blue, Lawrence Swan, my Jewish pal who lived round the corner from me, came up with a surprise proposal. He said: 'Why don't we go to Israel?' Lawrence said we could go to a kibbutz and could travel on a yearly return ticket with El Al, which would work out cheaper than going to Palma for two weeks. I said: 'Okay, fine!' The only thing I knew about a kibbutz was that it was some kind of a commune. Lawrence said many people who were not Jewish also lived there but not permanently. We were called 'volunteers', he said, we would be given a job and we would work. So off we went.

We landed in Tel Aviv and spent the first night there with someone his parents knew. The next day, we boarded a bus up to Beit Alfa kibbutz. An English chap, Paul Stevens, showed Lawrence and me to a hut where we were going to stay - an old army officers' hut from when the British were out there. It was sparse but pleasant. We both quickly got into the routine of the place. There were taps and sinks outside the huts for washing, where we met up with all kinds of volunteers from different backgrounds. Both of us quickly realised that kibbutzim are for fun, not just learning Hebrew. Around us was a bunch of eighteen, nineteen, twenty-year-olds, most of them from Sweden and Denmark, but some Americans and Australians too. We didn't get paid any salary but we got pocket money and they'd give us pretty much anything we wanted. If we wanted beer and vodka for a party, they'd give us it.

We all worked together, boys and girls in most jobs. All the children in the kibbutz lived together, not with their parents. The girls who weren't with the children worked in the kitchens and in the fields with the boys. The parents were always around and the children would go and have dinner with them, but they lived separately. We quickly made new friends. Each day we'd be taken in a tractor to the fields, where the day was spent picking grapefruit and oranges or putting posts in the ground. Occasionally, the Israelis would take us on a bus tour. There would be a coach full of volunteers and at the front and back there would be a couple of guys from the kibbutz, each carrying Uzi machine guns. Each driver would be driving with an Uzi slung over their shoulder too. To me it was fascinating and I thought: 'It isn't like this in Scotland!'

I stayed on Beit Alfa for a month, and when some of the volunteers decided they were moving on, I decided to go with them. The people running the kibbutz were not particularly chuffed that I had arrived, stayed a month and then just headed off. We went south and eventually arrived at another kibbutz called Hatzerim in the Negev desert. We decided to go down to Sinai and live in the desert, so we hitched towards the town of Eilat.

We met up with tribes of Bedouins and bartered with them for flour and trinkets, and would offer them clothing or old shoes in return. This was in occupied Egypt and, of course, the whole place was historically Arabic. At that stage, I was quite pro-Israel because I had been brought up to think of 'nasty Arabs' and poor old Israelis. But now we were starting to meet more and more Arab people who were always friendly and it was a very interesting experience.

It was around this time that I came across some girls standing around topless. I thought: 'Okay, that must be acceptable in Scandinavia!' It was the first topless lady this shy Glasgow lad had ever seen. We stayed in the desert camp for a few days, and one of the Danish girls was getting friendly with me. I found her quite attractive and when I decided I was going back up north to the old city of Jerusalem, she asked to come with me. We managed to find this vehicle called a sherut, a cross between a private car and a taxi, but by the time we got to Jerusalem, my new friend's constant chattering had got me to breaking point. I said to the driver suddenly: 'Stop!' I clambered out of the car, said goodbye and off I went into the old city alone.

I eventually found a place to stay called Mr A's, which I had been told about. It was like something out of The Arabian Nights. I got myself a job on one of the market stalls in the bazaar selling the usual tat to tourists. I had had an amazingly peaceful time in the desert, but back in the city, life was a lot different.

I befriended this Arab guy, who started telling me how they had such a hard time at the hands of the Israelis. I had already noticed that when he and I went out together, he would be stopped by the Israeli police, who checked his papers, but as soon as I pulled my British passport out, I was fine and allowed to move on. He started talking about how his grandparents had lived there all their lives and their land had been taken by the Israelis. I even persuaded him to come back with me to meet Lawrence at Beit Alfa, but it was around this time that I started to realise about the oppression of the Arabs. Spending time with this guy and going to his family house in Bethlehem, meeting other Arabs, I quickly understood they were no different to the Israelis. They just wanted to live their lives peacefully without oppression. It was a very different story to the one I carried in my head when I first went to the Middle East. They say travel broadens the mind, and I think it does.

But I was missing my flying. I just had to get back into the air. My plan was to go to Tel Aviv airport and persuade someone to let me fly one of the Cherokees there. I found a club and persuaded them to hire me an aeroplane with an instructor, and off we went. I flew for an hour around the Tel Aviv area and the coast and when I landed, I thought: 'Okay, now I'm going home. It's time to get on with my career.'

What I hadn't been aware of until I got back to the UK was that after I left that kibbutz, a British girl arrived at Beit Alfa. When Lawrence and I first arrived, two other British girls had just left, and we took their room when they returned to the UK. It wasn't until many months later that I found out that one of the two girls who had left, Jill, would turn out to be my future sister-in-law. And the girl who arrived after I'd left was to be my future wife, Ruth. So I met neither of them at that point. When Jill moved out of her room, she heard that two Scottish boys had moved in and thought: 'What a cheek.' It was her room. She told her sister Ruth the story when she got back to the UK, in Manchester where they had both been brought up. Little did any of us know that we would all be in the same family one day.

Dodgy dealer

I got a job back at the airport as a baggage handler, but now, rather than just a summer job, I decided to work there full-time. Of course, I could resume my flying with a little more money in my pocket. I got to know a flyer called John Fraser, who was about ten years older than me. He was quite a charismatic ladies' man and well-educated. He was a confident kind of chap and he was about to lead me wonderfully astray. When we flew together, it turned out that he was as big a hooligan as I was in an aeroplane. Soon we started hanging out together, and I started to get disapproving looks from the other pilots – 'do you know what you've fallen into?'

John and I started selling cars together, something I think he had done in the past, so I gave up baggage handling and became a dodgy car dealer. We rented the forecourt of a petrol station in Renfrew, south west of Glasgow, and for a while we did quite well. John knew a chap who had his own aeroplane, a car dealer in Glasgow who was a bit of a dodgy character as well. His son had dealerships, and they would sell us cars they didn't want which had probably arrived as trade-ins. Eventually, we rented a showroom in Charing Cross, Glasgow, opposite where my auntie's house would have been, on the other side of the motorway, before redevelopment. It was called Woodside Motors and we dealt in bread-and-butter cars and old bangers. Once a week, we'd put an advert in the newspaper saying 'Heap of the Week' and that car would have a 'Heap of the Week' sign on its roof. It was sold to anyone daft enough to come in and make us an offer. The newspaper thought it was a great advert.

We hadn't started with much money, just enough to buy our first car, and we slowly built up the stock of cars by buying and selling. Our common interest was, of course, flying, but John and I were caught up now in the business of trying to make money from cars and we'd go off to auctions looking for bargains and had a good time. I had become an accomplished drinker in Israel and about this time, the barman at the Glasgow Flying Club said: 'You are drinking too much, Scott.' He was Polish and was expressing concern that I just ignored. He had watched as I was gradually being led astray. He had also probably heard that John and I were doing more and more 'hooliganising' in aeroplanes. John and I would go up to the Isle of Mull to stay with David Howitt. We would spend time in the aeroplane flying around and attacking things, like motor cars or fishing boats. In Mull, no one really cared what you were up to – as long as you didn't crash the thing.

I eventually got fed up with our Del Boy car dealing antics. We weren't making any real money, so I said to John: 'I've had enough. It's your business. I don't want anything. It's yours.' I just walked away. John went bust some six months later, and I heard that he ended up driving taxis, although he did get his professional pilot's licence eventually.

My only ambition was to get back to flying. By 1975, now aged twenty, I had managed to arrange a part-share in an aeroplane acquired during my time in the car business. There were five of us, and together we found a Jodel D112, G-BAAW, an old two-seat, 65HP aeroplane, French-built with a canvas wing and very basic in all aspects. It turned out to be older than me. The fuel gauge was a rod in front of the pilot with a cork attached. When the fuel went in, it would float up and as the fuel went down, the cork went down. The trouble was, the cork would sometimes stick and then suddenly drop down to indicate considerably less fuel than I thought. You had to swing the propeller by hand to get it started. There were no electrics. We found this aeroplane in Sunderland at an airport which is now the site of a huge Nissan factory and bought it from a teacher for £1,200. I had a bit of money for this venture, squeezed from the car business, and the total purchase price was divvied up between the five of us. Robbie, the ex-military chap who flew the Aztecs, was asked to go to Sunderland to pick up our new acquisition, since he was the most experienced pilot, and I went with him.

Robbie got this thing airborne to fly back to Glasgow Airport. It didn't have a radio so we borrowed Loganair's ferry pack, a little battery radio thing which George had organised. We landed after about an hour's flying time. We put it in the Loganair hangar where George was chief engineer and, for about a month, it was worked on. Some of the canvas wing was re-covered, doped up and painted in a new colour

Robbie (right) and me arriving at Glasgow Flying Club with our Jodel

scheme. It ended up white and blue, and at the back they painted a little set of wings along with the name of our aeroplane – The Flying Scotsmen. We moved the Jodel to Strathaven, a tiny strip which had a windsock and a barn where we could store it, about an hour south west of Glasgow. It had to be moved there because landing charges at Glasgow were proving costly. And of course, we didn't have a radio. We had to borrow the ferry pack, which caused a bit of a problem on one occasion. We had got airborne, and the battery slid on to a terminal on the handheld radio, and suddenly there was smoke. We put out a PAN call (urgency call), and came back and landed, generally causing a bit of chaos.

Eventually, the authorities at Glasgow Airport said: 'Move it!' So we did. We each owned an equal share of the aeroplane, and to use it you simply booked it in the group diary. Very quickly, it was realised that John and I were flying it for most of the time. No one else was booking it on a regular basis, but John and I were flying it a lot. We'd fly around the area and take it over to Mull, clocking up the hours. It only burned about three gallons an hour, so it was very cheap to fly. I also took my brother Nigel flying in it. I didn't take Nigel up very often - at that stage he was not expressing any real interest in flying. He was about seventeen and was working for Woolworths. The canopy of the Jodel was a metal frame with a perspex screen which, one day when flying with Nigel, opened in mid-air, causing the perspex to crack. I landed in a farmer's field because I didn't know if the whole thing was about to come off. The farmer drove out, had a fiddle about, and it seemed all right so we just shut the thing and got airborne again.

One night, a fierce storm blew through the airstrip where the Jodel was housed. A large chunk of the hangar door was ripped off and crashed on to the top of our wooden propeller, smashing the tip. George had to come up with an inventive and low-cost solution. He simply cut off about two inches of the damaged end of the prop! To keep it balanced, he cut the same amount off the other end. From that point on, our propeller used to whizz round quite a bit quicker, but somehow it was still persuaded to do the job. These days, that propeller is hanging up as a trophy in my house.

All this was going on in an aeroplane where the air speed indicator was also a bit dodgy. It had an oil gauge marked 'Huile' but I never knew how to read it. I just made sure the needle was always in the green zone. The rev counter worked okay, but there was no artificial horizon. There was an altimeter, but nothing else. The throttle was a lever on the instrument panel.

Eventually, the group became a little disgruntled and said: 'This is ridiculous, no one is flying it, except you two.' I decided to buy them out. I paid £1,500 for the aeroplane, which Dad lent me. Since it had been worked on and improved, I felt it was a lovely little aeroplane. I bought everyone out, including John, because he and I were starting to part ways a bit. Even I was beginning to see that he was trouble.

The immediate problem was: 'Where do I keep this thing?' It was on a PFA (Popular Flying Association) permit to fly, which meant I couldn't fly over towns or take it out of the country, but I could do some of my own maintenance. I was pretty fastidious. George had always serviced it very well so I took this seriously because I wanted to keep it in good condition.

I decided to take the Jodel to Mull, where David Howitt had said I could live for a while. I was still occasionally flying a Victa Airtourer, a great little aerobatic aeroplane originally from Australia. Some were re-built at Staverton, Gloucester. It was a hot little aeroplane operated by the flying club at Glasgow with a 150HP engine and constant speed propeller. It had two engine control levers, one for RPM, and the other for boost. The control column was in the centre. It was fast, very manoeuvrable and was stressed to +6g, -3g. Robbie and I had attempted a little aerobatic flying, as he was quite experienced, but in next to no time I found myself getting airsick. I took it up a couple of times alone because I couldn't afford someone teaching me.

By now, I was the proud owner of a series of Birch and Bramson flying books, Volumes One, Two, Three and Four - and Volume Two included aerobatics. I took this book with me in the aeroplane, so I could learn techniques of aerobatic flying while I was actually at the controls. On the second occasion, I was over Loch Lomond, at about 4,000 feet, earnestly reading about how I could do a loop. I stuck the book under my leg, and put the aeroplane into a dive to prepare for my wild attempt at my first loop. I got to about halfway through the manoeuvre at the top of the climb, when I completely lost it. I was at the top of the loop and inverted and because I had lost any positive G, all this debris from the floor was flying about my ears. It's amazing how much stuff there is on the floor of a small aeroplane.

Somehow, I managed to get out of that and I thought: 'Okay, try a barrel roll this time.' I flew down and staggered around, then decided: 'Okay, one more go at a roll.' That ended up almost being the last command I ever gave myself. I got the aeroplane inverted and, just at the moment when the good book would have told me to reduce elevator pressure, my intuition took hold and told me to pull the stick back instead. Disaster! The water of the loch screamed towards me, out of control and gaining more and more speed. Somehow, I pulled out, just before I hit the water. Then, on the way back up, I was going so fast – this machine could go about 180 knots – I managed to execute a roll. It was not subtle – just bang one way, bang the other way, really badly flown. When I was straight and level again and regained my composure, I made another important decision. I took the book, threw it in the back and thought: 'Forget aerobatics!'

A little later on, and having recovered my nerves, I did decide to give aerobatics one more shot, and the location for my next attempt was again going to be over water, near Greenock, where the River Clyde turns. Once again, attempting a loop, I found myself coming down too low and too fast. I was desperately pulling hard to try not to hit the sea and then, to make matters worse, I started to 'grey out'. When this happens, you experience tunnel vision and just before you black out, you lose all peripheral vision. You just start staring down into this tunnel. Very scary. I might have been pulling in excess of 4G or 5G at that point, and I remember thinking: 'I'm not sure I'm going to make this.' I found enough sense to release enough back pressure on the control column to regain some control and came out of it thinking: 'Yes, I'm still alive!' That was it – I never did any aerobatics again. Since those early days, many aeroplane owners have offered to take me aerobatic flying and I have always said: 'No way!'

I first came across the unbelievable thrill of low-level flying on my first solo cross-country flight, over the waters of Loch Lomond. Suddenly here I was, let loose with no one sitting beside me, watching me. I could do what I liked! And I suddenly quite fancied going down and beating up the loch. Why? Because it just made me smile – and it still does. It was exhilarating and fun, but it needed some skill because at every point you had to think about how the aeroplane was trimmed. You didn't want to lose a speck of concentration. So I kept it trimmed to ensure that if I did lose concentration for a second, the nose would pitch up. I was thinking about my speed, thinking that when I banked, how close my wing tip would be to the water, and where I would go if, for some reason, the engine stopped suddenly. I had to be carrying quite a lot of forward speed so I would be able to pull up and then decide where to go. At that point I'd have very limited options.

It was a bit naughty, but not completely illegal. The rule in those days was that you couldn't fly any closer than 500 feet from any person, vessel, vehicle or structure. If I was skelping by a wee island at low level, that would have been completely legal, unless I came across a boat, and it was less than 500 feet below me or to the side. But my thinking was always that if I was going to go over a wee boat, no one on board was going to take my registration number – they'd be too busy diving for cover, wouldn't they?

I was always going to go as close as I dare, and I suppose that meant I would get down to twenty or thirty feet at times. Flying at speed at treetop height by the islands was exhilarating. The sense of speed going in among the islands, and then banking away, was sensational. Robbie would have taken me up, and done a bit of low-level stuff but, until that first solo flight, I had never been able to do it myself. I was too stupid to realise that at that point I would have had thirty or forty hours flying experience to my name, whereas Robbie would have had thousands of hours.

Time flies

When I look at my log book today, I can't believe how much went on in such a very short time. From leaving school to working as a baggage handler to going to Israel was all achieved in about eighteen months. Then, becoming a flying instructor, and getting married was all about to be achieved during the next eighteen months.

Within a year of coming back from Israel, I rented a cottage down by the airport with my friend, Lawrence. He was engaged to Anita Bock, a 'Swedish' girl who lived in Sweden but turned out to be Danish. He announced that the lovely Anita was coming to this country and she was going to stay with Ruth, who had come back to the UK from the kibbutz and was teaching drama at a secondary school in east London. The plan was for Anita to stay with Lawrence in our little cottage. As soon as she arrived, I got a sense that there was no great relationship between the two of them - they were more like friends. Anita turned out to be pretty, bright and fiercely confident. It didn't take long before I took a shine to her – and she took a shine to me. The next minute, we were going out.

Not unnaturally, Lawrence disappeared at that point and Anita stayed on with me. I took her to meet my parents in Giffnock, where we stayed for a couple of nights, and then we flew up to Mull in a Cherokee. She really liked flying and, later on, even achieved her Commercial Pilot's Licence. We stayed at the hotel and had a lovely time. Then, out of the blue, she announced she was heading off. I travelled with her by bus to London, where her plan was to catch the boat home from Tilbury. Anita's departure was about to trigger another landmark moment in my life. She was going to stay at a friend's flat on the night before her departure, so I stayed over with her. And that's where I met Ruth, my future wife, for the first time. Ruth rented the flat with another teacher. It was in Upton Park, east London, on the twenty-first of twenty-three storeys.

I stayed a couple of days. Nothing had been said, but I sensed a cooling off between Anita and me. Someone had warned me that Anita had broken many hearts and I had a feeling I was next. So the next day I put her on the boat with a clumsy kiss goodbye and turned away feeling confused. In pretty low spirits, I made my way back to the flat and was made very welcome, and told: 'You can come back and stay whenever you want.'

I went home to the little cottage and found myself moping around for a few days, trying to get over my newly broken heart. Not many days passed before I made another decision – I was somehow going to get to Sweden to try to rekindle my romance. She was living in place called Vilhelmina, just about on the Arctic Circle. This was early 1977 and I said to some pals: 'Why don't we fly to Sweden?' It was bound to be quite an adventure in any light aircraft.

Some years before, I had become friendly with a chap called Bill Wilson when I worked as a barman. I was eighteen and trying to earn money to pay for my flying. Bill was the junior partner in a motor repair business run from a hired barn. When I went off to work at the airport, I rather lost contact with him. Then I bumped into him one day and he told me that had learnt to fly. I wasn't greatly surprised because he was quite well off and had been banned from driving. He had used his ban to seek other forms of mobile recreation, and had somehow found his way to the West of Scotland Flying Club, the other club in the Glasgow area, where he learned to fly and got his PPL. He later went on to achieve his professional licence.

We both enjoyed flying and he and I became pals. So when the time came to think about flying off to Scandinavia to find my lost love Anita, Bill was one of the people I asked to accompany me. Before that, I wanted to take advantage of the invitation to return to the flat, so Bill and I drove down and stayed over a few times - and that's when I started to notice Ruth. I was still quite struck on Anita and it was going to be sometime before another girl attracted my interest.

Bill had no ties back in Scotland so I had no hesitation in asking him to join my crew for our madcap adventure to Sweden. The other chap I asked was a PPL, but not very confident in an aeroplane. He had got through his training without a problem, but flying never really came naturally to him. Nevertheless, importantly, he agreed to pay his share of the cost. When I said to them 'Let's take a Cherokee and go to the Arctic Circle – it can't be that difficult!', both of them had, perhaps foolishly, agreed. It was summer, and the plan was to fly northwards towards Inverness, but the weather wasn't looking very good. Shortly after leaving Glasgow, we thought: 'We haven't even managed to leave Scotland yet. Why don't we go down to Newcastle and cross there?'

That leg represented a long crossing of the North Sea with just a single-engine aeroplane. Nevertheless, we filed the flight plan with ATC (air traffic control) at Newcastle – they didn't give our idea a second glance. They probably thought we knew what we were doing. We cleared customs and off we went, straight across from Newcastle to somewhere on the west coast of Denmark – quite a few hours over water. We had a VOR navigation system, so we flew out on the VOR beacon from Newcastle but, at our height, it was only going to be accurate for about fifty or sixty miles. After that, we just kept going. I was too stupid to realise just how idiotic that was. If we had ditched, we were all gone. Eventually we were heading into the airfield on the Denmark coast and there was no beacon on it. To make things worse, we were in cloud.

I had an IMC rating so I was trained to fly in cloud, but not to carry out proper instrument landings – and it was only valid in the UK. The best I could do was to stumble around with the little knowledge I did have - so we were not necessarily going to die immediately. By our estimate, we should have been approaching the airfield, but we were in cloud and just didn't know our exact position. Radar informed us we had already overflown the airfield.

It was not an impressive situation. There was me, a new private pilot, with not much experience, and there was a radar controller saying that we had to turn, and me requesting: 'Say again.' And them replying: 'Turn, confirm.' That didn't seem right. Fortunately, at that point, there was a gap in the cloud, and I could see what I thought was the sea. I had the map out, thinking: 'Have we really overflown the airfield, or are we now looking at an inland lake?' It turned out that the controller had got confused, and the bearing he gave us was 180 degrees out. If I had followed his instruction, we would have ended up in the North Sea. That was the first time I had nearly been killed by air traffic control – but it wasn't to be the last. We landed eventually, but when the chap filled us up on the ground he commented that our tanks were almost dry. At least we were one flight closer to our destination.

We flew across to Copenhagen in Denmark, then flew north. We had a map which wasn't always very informative, particularly when it described the ground below as having 'Lakes too numerous to plot'. On one leg, we found a wonderful little airfield with very good tarmac and a tiny terminal - a gorgeous little place. Only problem was, there was no one there. We had no option but to take off again, and we eventually got to Vilhelmina. My spirits were raised by the sight of the lovely Anita at the airfield waiting to greet us. Sadly, any hopes of rekindling my romance were quickly dashed when she casually announced that she was now going out with a new local chap.

That Swedish trip was typical of the kind of aviation I was involved in back then. Every time I got airborne, it was going to turn into an adventure. We always seemed to be pushing the limits. I wrote an article about our Swedish adventure for Pilot magazine when I got back. One reader was moved to comment: 'Well, that was a bit daft!' On our way back, we went across to Norway, and the next stupid thing we did was launch from there hoping to get to the Shetlands. Again, we flew out on a VOR, but we were on top of the cloud. We lost the VOR because we, again, ran out of range and I remember thinking: 'How do we find this island?' We hadn't quite considered how we were going to find Sumburgh Airport. The cloud was flat except for a small bump, and I reasoned: 'It's got to be under that.' So we spiralled down a few miles short of the bump and emerged out of the bottom of the cloud. The island was, indeed, under the bump of the cloud, so we flew in low-level under it. We landed, refuelled and flew on to Wick, Inverness, then down the Great Glen and back to Glasgow. We had been away a week or ten days and it was the longest flight I had taken. The three of us had had a fantastic adventure, and I had gained valuable hours towards my instructor's rating.

Mulling over another near-miss

By now I had resigned myself to the sad truth that the lovely Anita was out of my life for good. So it was pleasant to consider that I would be welcomed by female company down in London should I wish to stay at the flat at any time - a thought which brought some comfort as I nursed my way through my first romantic rejection.

On Mull, by that time, I had come to be called The Pilot - I'd walk into a pub in Tobermory and someone would announce: 'It's the pilot!' I had a lot of fun there with Bill. David was building his own house in the grounds of the hotel at that time, and I would help him work on the house for a while, before taking to the air again to fly my own aeroplane around the island and over to Oban. David was quite a dab hand at flying little radio-controlled model aeroplanes, and it became one of our favourite pastimes to try to fly the Jodel and his little radio-controlled model in some kind of formation. We'd both take off in my aeroplane, and David would attempt to control his model plane, flying alongside the cockpit. It wasn't a great success. As soon as I began a turn, that would be the end of our synchronised flying. The little model plane would just be gone.

I had moved up to Mull permanently while all of this was going on. I didn't really have much with me, just a few clothes and an aeroplane. David and I would go on fabulous walks up the glen. A Dakota had crashed there during the war, and we looked for part of the wreckage. We found a big radial engine by a riverbank. The crew had been ferrying across from the USA, on a delivery flight, and tried to get down below the clouds but hit the hillside.

With my Jodel, I was quickly learning that some standard manoeuvres could be a bit tricky. Because it was a tailwheel aeroplane, when landing in a field, you'd try to do a three-point landing. On a hard surface, like a tarmac runway, if there was space available, you'd tend to land the main wheels and then, as you reduced speed, the tail would just sit back down. I had to remind myself that no one had actually taught me how to fly it. I had just picked it up. There are four main forces which make a single-propeller aircraft with a tailwheel turn to the side of the runway on take-off. If you were in a powerful aircraft, such as a Spitfire or Mustang, and you opened the throttle quickly, you would instantly disappear off the side of the runway. The effects are gyroscopic effect, asymmetric blade effect, torque effect and slipstream effect. All four act in the same direction, and with a propeller rotating clockwise as viewed from the pilot, you will get a swing to the left which requires right rudder. From the cockpit you can feel and see the aircraft starting to turn one way. This is a big challenge in a powerful aeroplane like a Spitfire, but with the little Jodel with a 65HP engine, the force wasn't that powerful. That's why I didn't kill myself. I just stumbled around until I got the hang of it. I didn't even know that I was meant to push the stick forward on take-off. The correct technique was to keep the stick back and then, once you'd started to put power on and gained some speed, you'd push the stick forward to raise the tail. Then at flying speed you'd rotate. Now, I didn't know any of that to start with. I just kept the stick back until I sort of mushed into the air. Because the Jodel wasn't a powerful aircraft, I didn't get a lot of swing.

Anyway, now the Jodel was on Mull, it didn't matter that there were no electrics and no radio. When I flew it into Glasgow, I borrowed a battery pack and radio. The engine had to be started by swinging the propeller. It was only 65HP so it didn't require a strong arm to swing it, and it started easily. By the time it arrived on Mull, one of the heel brakes had failed, which meant that on the ground I could only really turn it one way. The airspeed indicator had stopped working, and all that did work were the oil gauge, RPM gauge and the cork float measuring the fuel. But other than that, it was fine!

I didn't really need an airspeed indicator because I could feel when I was flying slowly as the aeroplane would start to get a bit waffley. It was proving to be inexpensive to fly. My bottom would get sore before it ran out of fuel. That meant I could go and play for an hour or more around the islands. I did do a couple of longish flights, and after a while I began to think: 'It's not going very fast.' That didn't matter too much when it came to having fun.

What I really used to enjoy doing was flying down low and bouncing my wheels along a quiet country road somewhere. There were little single-track roads up in the hills, and also on the way to Tobermory on the main road. If we were larking about out at sea, Bill and I always used to say that the only time we could be sure the boat we were attacking was not a Russian trawler was when we could fly so close that we could see what the skipper of the trawler was reading. Was he reading an English paper or a Russian paper? The only way to find out was to fly low and at close quarters. We would quite often attempt to fly through the gap between the bridge and the mast, just for the fun of it. The only thing I failed to do was to spin my wheels along the surface of the sea. I always chickened out. But on land, we would never go over a pylon; we would always fly under it. All the time, we were having a ball, simply because there was no one there to say: 'Stop it!'

Our antics around Mull didn't go entirely unnoticed. Because I was making a habit of low-flying around the island, quite a few times I landed back at the hotel and someone would say: 'The police have been on the phone.' I'd have to phone them up and talk to the local bobby Alec McGlashan, and I would say: 'Could I speak to Constable McGlashan, it's Scott Birrell.' And he'd say: 'Oh, hello, Scott.' And I'd say: 'I wasn't any closer than 500 feet to any person, vessel, vehicle or structure.' He would say: 'Och, I'm not worried about that. Someone said they had seen you crashing, so I'm just checking that you are okay.' I'd say: 'I'm okay, Alec. Thank you very much.' I would probably have been attacking something but hadn't actually crashed, although from ground level it probably looked as if that would have been the only outcome. Someone might have seen me disappearing over a hill at ten feet or so.

I was flying along over the island one day, minding my own business, when bits of my wooden propeller - shortened after the mishap with the barn door - began to break off and fly off all over the place. When I landed, the propeller looked like it had been through the wars. The outer hardened layer which covered all the wood was tearing and breaking off. I thought: 'This isn't good.' So I called the PFA, the Popular Flying Association, and got through to a chap who was advertising propellers for sale. I said I had a Jodel D112, and he said he could make me a propeller but it would take a month. So the aircraft was grounded because there was no way it was going to be safe to fly. I took the old prop off, which in hindsight I'm not sure I should have done, and we tied the aeroplane down, parking it behind the old fire engine, which we used as a windbreak.

When it was grounded, I topped the Jodel up with fuel because I didn't want water to get into the tank. With something like a Cherokee you can easily drain the fuel from the tanks under the wing to get any water out. Much later, when we had to do that with the old B737s, we would end up draining as much as half a litre of water out of the tanks. The little Jodel didn't have that; it had a large sump drain. I reminded myself: 'When I get the prop done, I'm going to get it back to Glasgow and the engineers can fit a more convenient drain.'

So I left the Jodel full of fuel and there it sat for a month. Eventually, the new prop arrived, and of course immediately I saw it was bigger than the modified old one. I bolted it on, torqued it up – although I'm not sure I should have done that either – and I thought I had better take the aeroplane up and try it out before I took someone flying in it. The whole crowd was there as I was due to take some of them flying later on. It started on the first swing of the propeller and off I taxied. I noticed there were sheep on the grass runway, but that was not uncommon and wasn't a problem. I simply taxied up and down to get rid of the sheep as I had done many times before, and off I went. I took off and flew over the bay. The aircraft felt absolutely fine. The first thing I noticed was that it wasn't revving as much on take-off because the new prop was now absorbing the power, as it should.

As I came in to land, I noticed that the sheep had walked back on to the strip. That wasn't really a problem. I just thought: 'I'll do a low buzz and if the sheep don't move, the crowd watching me from the hotel will know, and they'll go out and move them.' Just as I pulled up, the engine faltered and then stopped completely. I didn't know it at the time but there was water in the fuel pump, feeding through to the carb. To compound the problem, my airspeed indicator wasn't working. So I lowered the nose and thought: 'If I can get into the field next to the airfield, I should be okay.' As I neared the end of the airstrip, I could feel the ailerons were getting very mushy, as the airflow was slowing over the wing. At the end of the airstrip was a typical wood and wire fence, and I knew I needed to clear it to get safely into the field beyond. As I was getting closer and closer to the fence, flying quite low without any power, I could feel that if it stalled and dropped a wing, I was going to crash and roll the aeroplane. It was going to be ugly. So I had to lower the nose and part of the aeroplane caught the fence as I skimmed over it. I just couldn't maintain enough height. I must have been flying very slowly, and then, having caught part of the main fuselage on the fence, it just rolled over on to its nose.

I was probably doing about forty miles per hour into the wind and even slower over the ground. But at least I was in one piece. Just a little shaken but unhurt, I hurriedly turned off the fuel and scrambled out.

I was further relieved to find that my brand new prop wasn't smashed to bits. It was marked, but not badly damaged because it had stopped horizontally.

I was soon aware of the crowd of onlookers I had been trying to impress, all running down the strip to see if I was okay. I was about 300 or 400 metres from the hotel. I had bashed my leg a bit, but not badly, and there were no broken bones. In fact, I had come out of the wreck of my Mini van in worse shape after I wrote that off some years earlier.

Back at the hotel, more than a little embarrassed, I thought I'd better phone the police once again – in accordance with the rules of flying – because I had crashed an aeroplane. So I phoned the local police and once again spoke to PC McGlashan. Alec said: 'I'll be down shortly.' I phoned the CAA in London, again according to the rules, and said I'd had a plane crash. 'Was anybody hurt?' 'No.' 'Was there any damage?' 'Not really, just to the aeroplane.' They said they'd phone me back, and ten minutes later they said: 'Fine, carry on!'

Then the police arrived, Alec driving his Morris 1000 with a blue light on top. He picked me up and we drove soberly down to the strip, past the hotel, to the scene of the crime. There was my little aeroplane, sticking up out of the ground. Instead of driving straight to my aircraft, he drove down to the strip, and then began driving along the grass runway, saying: 'Right, we will be the aeroplane.' Obviously, the highlight of this country bobby's year was me crashing. 'Now, Scott,' said Alec, on arriving at the scene, 'I want you to tell me, using all the big technical words, just what happened so I will impress my bosses in Oban.' So I told him: 'It may have been carburettor icing, I don't know.' 'Oh, marvellous,' said he, writing it down in his notebook.

Then the party began. Having treated my modest wounds, my friends and a few locals carted me off to the nearby bar in Salen, and got me mind-numbingly drunk. If I had felt like lying down in a darkened room and thanking my lucky stars, it was never going to be an option. Having captured me in the bar, they locked all the doors to prevent any escape.

But this partying was to lead indirectly to another casualty which upset me more than my own physical injuries. At that time, I was the proud owner of a jacket – a real World War Two leather flying jacket. It was given to me by a man at the flying club who also donated pilot boots and big heavy dungarees from the Second World War, in recognition of my exploits in the air as a teenager. It was a gorgeous jacket and I was wearing it at the time of the prang, and throughout the revelry which followed. This had been the first real prang of my career and now I was absolutely wasted. At the end of the evening, when I tried to stand up, I fell forward and went straight through a table with a loud crash. I had to crawl out of the debris in my lovely jacket, which was by now of course the worse for wear.

I don't know how much vodka I drank, but I have never touched the stuff since.

An American adventure

Bloodied, but not the least bowed, I went back to Glasgow and started flying again at the club. Now I had to concentrate on building up my hours. At the time I crashed the aeroplane – a ten-minute flight on April 18, 1976, according to my log book – I had a total of 149 hours, of which 98 and a half were P1 (Pilot in Command). So I was just 50 hours short of my target to go on an assistant flying instructor's course. I continued to fly with Bill and we'd occasionally go back to Mull in the little Airtourer 150S (registration G-AZZU) but I was back baggage-handling for a while. I was also doing flights to Strathallan, in the middle of Scotland. Basically, I was doing anything I could. All the time, as I was building up hours, I was getting my log book signed every month to make sure all was correct when it had to be checked by the CAA. I took advantage of my time at the controls to practise short-field take-offs when the opportunity arose. I was taking people on some of these flights and they were chipping in towards the cost. If the cost per hour was £50, they would chuck in a tenner each, so it meant I was getting cheap flying time. The Cherokee 140 was never comfortable with four on board. I remember dragging the aircraft off the ground with four big people to go to the Isle of Man, and it was only really the curvature of the earth that got us airborne.

Then, that May, something amazing happened. I had just logged up my 106th hour P1 and was at the Flying Club. I happened to overhear the conversation of a chap called Jock Adams, a British Caledonian BAC 1-11 captain who was generally known to be a bit of a troublemaker and a real handful. He used to come to the club when he arrived at Glasgow and one day he came in and said he was now doing ferry flights. He'd fly anything anywhere, and announced: 'I'm going to take a Piper Aztec, a six-seat twin-engine aeroplane, to America. Does anyone want to come?' Without pausing for a moment to consider what I might be letting myself into, I declared: 'Yes!' He immediately began briefing me, then a major obstacle presented itself when I remembered I had no money.

Behind the bar at the Glasgow Flying Club was a vast collection of notes of different currencies, including quite a few American dollars, pinned to the wall. They were no doubt donations from pilots who had flown in from all corners of the world over the years. To my great delight, one of the Flying Club committee members agreed that this haul of cash should be taken down from the wall and given to me. I still didn't really have enough for the return trip, but that was not going to stop me.

The deal Jock had negotiated was that we would fly the Aztec to the United States and I would then come back to the UK as a passenger on a scheduled airline. It all sounded straightforward. Jock would be delivering an aircraft which had been cloud-seeding in Africa before arriving in the UK. It was quite an old Aztec, and its most unusual features were ferry tanks to increase the aircraft's range, which had been fitted where the back seats had been taken out. As a result, we would be sitting quite tightly at the front with no room for the seats to go back.

On the day of departure, I went out to the aeroplane with Jock and he announced that we were going to go to Iceland, to Reykjavik, on our first leg. Then he ordered: 'Jump into the left-hand seat.' This was a real shock for no other reason than I had never flown a twin-engine in my life. I did know that as long as both engines were running, it was not hugely different to a single-engine aeroplane, but if one engine ever stopped, the thing would begin to yaw, and then I would need some skill to keep it the right way up. So off we went to Reykjavik, a smooth flight with no dramas, and we stopped there and refuelled. The Aztec had a nose wheel and double props, and I quickly discovered that providing I fed in the power on both engines on take-off, all was well. The right engine had a habit of stopping, the result of fuel starvation from the ferry tanks, but with a bit of fiddling around, Jock would eventually get it going again by hand-cranking the pump to get the fuel moving by turning around in his seat. The problem, we believed, was caused by an air block somewhere in the fuel system.

The flight time to Reykjavik was four hours and fifty minutes, and the plan then was to fly on to a military airfield at Goose Bay in Canada. That leg would take another eight hours and forty minutes. This was May so we were flying in daylight almost the entire distance. On the

way there I saw my first iceberg. We weren't flying that high because the Aztec wasn't pressurised.
Our flight path took us just south of Greenland, which has some very high mountains. We passed by the southern tip where there is a strong navigation beacon called an NDB, and it's still there now. There was also another system the U-boats used during the Second World War and aircraft used it afterwards. The Germans called it Elektra, and after the war the British used it under the name Consol.
Although I had flown most of the way, Jock landed it in Goose and we spent the night there. During the evening I strolled over to one of the hangars and came across a huge single-engine de Havilland Otter. A Twin Otter is a big aeroplane with a high wing and turbine engines, and this single Otter had one huge radial engine and was on floats with wheels attached. It was the first float plane I'd ever seen and I was immediately impressed by its size. I climbed up and found my way to the cockpit, only to discover that apart from an oil gauge, an ASI and an RPM gauge, it had no instruments. I found someone who worked at the airfield and asked: 'Why doesn't it have instruments?' He replied: 'You don't want the pilots going into cloud, no sir. If they can't fly, they land and put it in a lake.'
Next morning, just after take-off, the engine stopped again. We were heading for Stapleton in the direction of Chicago, and Jock asked: 'What would you do now?' I said: 'I'd turn back.' He said: 'Correct – but we're not going to… Keep going!' So we just carried on. I was now getting good practice with control on one engine, and once again he managed to get the rogue engine going again. We stopped and spent the night at Stapleton.
It quickly became evident to me that Jock had this unhappy knack of falling out with people. One time he had a shouting match with the Met man during our briefing, and everywhere he went there would be aggro. This didn't stop him being great fun to be with, but you couldn't ever really relax with him.
The next day, mercifully, we were able to leave, and we headed for Denver. We both had headsets and he got me to do the radio – which I found a bit amazing – while he was navigating between the VOR beacons.
Then we came to a big hole in the cloud cover. The Aztec was an American-registered aeroplane, but I didn't even know if he had an American licence. He ordered: 'Cancel IFR (instrument flight rules), we're going low-level' - no doubt because it was going to be cheaper. I was trying to get a word in to cancel IFR, but I didn't really know how.

He just ignored this and spiralled down so that we were coming over the central plains at low-level – very, very exciting stuff. Because it had two engines, the Aztec had a bit more of a heavy feel to it than a single-engine aeroplane, but the right engine was still playing up. As we were about to land at the old airport at Denver, the right engine stopped altogether and Jock said: 'Just leave it.' As we touched down, the left engine stopped too. Now we were sitting there with two dead engines, not going anywhere, and they were yelling in my headset: 'Clear the runway!' Eventually, somehow, we got one of the engines going and taxied in.

We were flying with a whole set of Jeppesen navigation charts which at the time I thought were probably more valuable than the aeroplane we were flying. It had turned out to be an old dog, really. In the back, you could still see remnants of equipment they used for cloud-seeding in Africa and tubes along the rear of the wing where the fluid came out. Anyway, Denver, thankfully, was to be our destination, and this is where we met the owner of the aeroplane. He was as happy to have his aeroplane back as we were to give it to him. To show his gratitude, he took us out for a lovely meal that evening.

Back at the hotel in Denver, we received a phone call. I was expecting to fly home the next day, although I hadn't a clue how much this was going to cost. I had some money but I'd no idea if it was enough. After the phone call, Jock announced: 'We've got a ride home but we need to get to Miami tomorrow.' By now, it was dawning on me that I knew absolutely nothing about the people we were involved with here. I didn't know who actually hired Jock. It might have been the aeroplane owner or it might have been a company, but it hardly seemed to matter as long as I could get home safely.

The next day, we boarded an American Airlines Boeing 727, the one with three engines at the back, and headed for Miami. The plan was to tech stop somewhere en route, and when we landed there, we stayed on the aeroplane. Jock decided it would be educational for his young charge to pay a visit to the flight deck. There was no one there, so Jock put me in the co-pilot's seat. Needless to say, the pilot arrived back almost immediately and was pretty unimpressed. This was another opportunity for Jock to show off his insane lack of diplomacy and he barked back: 'Oh, for goodness sake, I'm just showing the young man. And I'm a captain.'

We landed in Miami but our flight the next day was to be from a smaller airfield. When we got there, I discovered that we were going to fly an aircraft home. But this was to be no ordinary aeroplane. We were due to ferry back a completely re-equipped, luxuriously finished DC-3, fitted out as a private aeroplane. The story was that the aeroplane was eventually going to end up in Khartoum where the two Niles meet. It was white and beautifully painted, and the interior adorned with leather seats laid out in a lounge format.

The owner of this flying palace was due to accompany us, along with his newly qualified first officer. I quickly realised that I was not even there to fly it. Jock was going to fly it. Uppermost in my mind was the wish that Jock would at least keep a lid on things until we got home. I had seen him cause trouble, and by now I was getting the impression that he enjoyed it. Off we went, and after thirty minutes, out came the first officer, with his three gold bars on his shoulders, and he said: 'Jock wants you up on the flight deck.' I went forward and found Jock mumbling something about 'incompetence' and 'useless bastard'. He told me to get into the right-hand seat and said: 'He won't be back.' At one point, I was in the left-hand seat and alone on the flight deck. The rest were playing cards and gambling down the back.

We were flying over New York in good visual conditions. Then the left engine failed. The aircraft started to bank and roll to the left and the nose dropped. I pulled back and applied right rudder, as I had learned to do on the Aztec on the way out. Then I could feel the trim change as Jock charged up to the front. He didn't get into a seat. I guess he changed fuel tanks, restarted the engine and with a comment 'Carry on', went back to his card game. This aircraft didn't have ferry tanks, so it didn't have the range to fly all the way to Reykjavik, and we had to land at Narsarsuaq in Greenland, a remote place established as an airbase in 1941 for US aircraft hunting German submarines.

Pilots would have to go into Narsarsuaq by flying up the Tunulliarfik Fjord. There are two fjords in close proximity, and if you went up the wrong one, you were in very big trouble because it narrowed towards the end.Turning in the narrow space between the hills might not be possible and you would not be able to outclimb the hills. So the correct approach, I was told, was to fly up low level, and as you turned up the fjord, there would be the wreck of a ship which you could spot. That way, you knew you were over the right stretch of water. Narsarsuaq itself was at the top of the fjord and you had to land in towards a glacier which you couldn't outclimb if things went wrong. There was just one way in, and you would have to crank it round very sharply for a go-around. So we were flying in very carefully, and as we began the approach, everyone was on the flight deck checking that we were up the correct fjord and marvelling at the view. It was a bit like a bomber raid.

Then someone cried out: 'Look, there's the wreck!' We landed without drama, but we seemed to have stopped in the middle of nowhere, and we ended up staying there for a day or two, due to weather. We had to gain enough altitude to fly over that high ground, up to about 10,000 feet, and the weather wasn't up to it. Jock was carefully monitoring the Met reports, looking for a window in the weather, so we could cross Greenland in good conditions.

On take-off from there, we would have to climb quite high and icing could be a problem, so we needed to find a benign weather pattern. Once we were past Greenland, we could then drop down for the crossing to Iceland. We would have the fuel to get there but there wouldn't be much to spare. After a couple of days, the weather cleared sufficiently for us to get airborne, with a good tailwind at cruising altitude. We stopped in Reykjavik, and the next stop was Prestwick in Scotland, where I was to disembark. Jock and the others were to fly on and deliver the aeroplane to Khartoum, where it was going to be operated as a VIP aeroplane. At the time, it was the only DC-3 I had come across fitted out for VIPs in that way. It was nicely done, and had a special bubble-like window fitted on either side of the passenger cabin, which stuck out into the airstream so passengers could stick their heads out a little bit to get a better view. It would certainly have slowed down the aeroplane in the air, but DC-3s were slow anyway – they'd only be doing about 180 knots. The conversion had been done in Miami. I heard much later that the company had gone bust, which in aviation is completely normal. There is a saying: 'If you've got a lot of money to invest in aviation, you will come out with a lot less.'

Before I got my bag and headed home, it was time to say my goodbyes to Jock. If I had expected a grand send-off, I was disappointed. He thanked me, shook my hand, turned his back and was gone. And that was it. This was at the end of May and I felt content that in a few days I would be flying again. What I didn't know was that this latest escapade was about to be trumped – and once again Jock would be at the heart of the adventure.

Out of Africa

I arrived home from that trip just a few weeks older but a great deal wiser, and immediately set about the business of trying to build up my hours. I was now getting to around 130 hours P1, so everything was coming along nicely. Less than a month after I arrived back in Scotland, Jock, freshly returned from Sudan, contacted me again. 'There's another job – are you interested?' My immediate thought was: 'Am I just a cheap co-pilot for him?' But was I interested? I sure was!

On the American trip we had been pretty well ignored everywhere we went, which made our passage a whole lot easier. I'm sure someone did check our passports occasionally, but the overall situation wasn't like it is today, where the first thing you have to think about is security and procedures. Travelling was much less complicated. I didn't have a visa for the US and the visa waiver programme, if it was even running then, was only available for certain airlines. I had no idea about any of these things. If Jock went, I was there, and I never asked any questions. But for this second, thus far undefined trip, I was told I had to get another passport. Jock said: 'They're making a film called A Bridge Too Far with Sean Connery.' It was about Monty's plan of an airborne attack on the Netherlands, which was expected to end the war by Christmas 1944, and which finally failed at Arnhem. The plan was that we were going to pick up some Dakotas and fly them to a location for action scenes. I immediately thought: 'I can just see Sean Connery and me at the bar!'

So when Jock asked if I had an Israeli stamp in my passport, he said: 'Then you are going to have to get a new passport.' In those days, you could just whizz into Glasgow with a photograph to sort out a new passport. I was wearing these photochromic glasses which changed when light or dark, and the person behind the desk asked: 'Do you need them?' I said 'Yes', even though that wasn't true, and the picture in my passport ended up with me looking like a member of the Mafia. The plan was that I should meet Jock in London. I flew down a few days later with BEA, again with practically no money but nevertheless good to go on my next unknown adventure. Jock took me aside at Heathrow and said: 'You need to know a few things. We are going to Uganda, Entebbe, and they don't know what's happening out there. They've no idea, but we're going to pick up these aircraft, five Dakotas, and we're going to be meeting up with other pilots who have come in from all around the world.' My reaction was that this had all the hallmarks of another of Jock's ragtag schemes. Was I still interested? I sure was!

We were flying to Africa in the back of a cargo plane, a Ugandan Airways B707. Jock confided: 'You've got to be a bit careful because there are going to be other pilots with us, and there will also be people on board from a Swiss company who are involved in armaments.' I listened and replied, as coolly as I could under the circumstances: 'Okay, that's fine.' At the last minute, there was a flurry of phone calls and Jock announced a change of plan: we were now going out of Stansted. So we both found our way there and, in a hotel room, met up with the other pilots. They were a real mixed bag and I thought: 'I'm glad I'm flying with Jock. I don't even have a twin-rating.' There were a couple of other people at the edge of the group who said next to nothing, and my guess was that these were the people from the shadowy arms company. Strangely, no one seemed as excited as me at the prospect of drinking at the bar with James Bond.

As we began the take-off sitting in the back of this cargo aeroplane, Jock asked: 'How long does it take a B707 to get airborne?' I said: 'I've no idea, twenty seconds, something like that.' This thing rolled and rolled and rolled and at one point it felt like the pilot was going to try to drive it to Africa.

We arrived in Entebbe, to be given a final reminder from Jock that what we were doing was secret. Even at that stage I didn't know any of these men, but I did know that it was prudent not to ask. At the back of the aeroplane, a pair of steps arrived. The rear door swung open and a Ugandan guy standing at the top of the steps immediately asked: 'Are you the guys who have come to get the DC-3s?' Now I was confused. We had just been told this was all completely hush-hush and now even the people hauling the steps at the airfield knew about it.

When we went to inspect our charges, Jock disclosed that these particular aeroplanes hadn't been flown for about five years, but he assured us they had all been thoroughly checked. Each turned out to be camouflage-painted, and down the length of the fuselage there were two-inch holes in each window from which a rifle could be fired. The floor of the cabin was bare metal, and there was a spare engine on each aircraft as cargo, located near the front. The rest of the interior was empty. They were basic military aircraft. I thought: 'Okay, this makes sense. They are going to help make a war film called A Bridge Too Far, after all.' As I said that to myself, it dawned on me: 'I'm going to be in the film. What a great idea.' There were double cargo doors at the rear of the aircraft. Jock said one of the five Pratt and Witney radial engine aircraft had two new engines fitted, and I asked if that was the one we were going to take. He said: 'Good god, no! You don't want to fly an aircraft with two new radial engines.' It was decided that an American chap was going to go up to test the aeroplane he'd be flying. I thought: 'I'll go along for the ride.' There was a co-pilot with him on the flight deck, and me. We got airborne from Entebbe and soon passed over a big lake, possibly Lake Victoria. We were flying quite low along the lake and the pilot was testing things, taking the aeroplane close to a stall. I thought: 'He's flying the wrong speeds.' Even with my limited experience, I knew what speeds he had to fly at. This aircraft was rated in mph, which was a bit unusual, but the pilot was flying as if the reading was in knots. So he was roughly 15 per cent out! Luckily, we survived. All the aeroplanes had radios but only one had a VOR, and one had an ADF (automatic direction finder). Between them all, they would have made one reasonable aeroplane.

At the airfield, I wandered into a hangar and was promptly kicked out by a bunch of military security men. But not before I spotted a gleaming Gulfstream, a fancy private jet, which turned out to be the prized possession of none other than president Idi Amin himself. I was told: 'That is always filled up, ready to go. And there's always gold on board.'

We held a meeting in the hotel and the discussion was about what would happen if one of the aeroplanes went down. The idea was that some people would fly on and some of the aircraft would circle and hold their position. I thought: 'How exciting, someone going down in the African bush.' On the day of our departure, we all knew the plan. We were going to fly up to Khartoum, and then on to a destination, as yet undisclosed. So we all got airborne, and the plan was that we would attempt a formation take-off. Jock and I were the lead aircraft and I was in the right-hand seat, but it all ended up a bit messy and there were aeroplanes all over the place. In the air, we were in a 'V' formation, Jock and me leading. The chap beside me, just off my right wing, was a hugely experienced pilot and his aeroplane was tucked in really tight. I was impressed. I was pretty sure he was flying like that just because he could – he was that good. I felt I could almost walk across to his aeroplane, so close were the wing tips. In stark contrast, the other aeroplanes were really floating about. There was no pressurisation so we couldn't go above 10,000 feet, but we were getting great views as we headed north to Khartoum.

On arrival, we caused chaos because our aircraft were going all over the place as we tried a formation landing. We weren't planning to stay, just refuel, because our next destination had been identified as Cairo. No sooner had that plan been adopted when Jock announced that no, we were not going to Cairo – 'it's a pain in the arse there; we are going to Luxor'. By now, I didn't really know where all this was heading – I didn't care as long as it was leading me to a bar and Sean Connery and Martinis somewhere! Since then, every time I have flown into Luxor, I remembered all this with great affection. When we did all manage to get our DC-3s on the ground at Luxor, there was a problem. To enter the country from the south, as we were doing, we all needed a yellow fever jab. I didn't have one and neither did half the chaps I was with. So while those in the clear went off to a hotel for drinks and a jolly good time, the rest of us were carted off in a vehicle, under guard, and driven to the nearest town, where we were banged up for the night.

This was another new experience. It was the first time I had found myself in jail. My incarceration was more a quarantine than jail – although, in countless bars around the world in the years since, I've always called it jail because it makes for a better story. Either way, we were still locked up. We couldn't escape because there was a soldier with a gun, an old Lee Enfield rifle, standing guard. At one point, we were allowed to sit outside our cell and the guard handed his rifle to the American, who was showing off impressive spinning tricks with this gun, the kind you see in American parades. All in all, it didn't seem terribly threatening – but that was about to change.

Next morning, I was lying on my stomach on the hard bed in our little cell wearing just a wee pair of underpants. I wasn't fully awake when I felt a hand, a male hand, running up my leg, over my bottom and up my back. In fright, I sat up sharply, and there, standing over me, was this huge Arab. With a leery smile, he handed me a single flower. I feared the worst and sat there, frozen in a cold sweat - but he didn't do anything. He handed me the flower, turned on his heels and walked out. It was only when he had gone that I noticed the other two chaps in the cell were also sitting bolt upright, in shock, each holding a small flower. We all agreed: 'It's time we weren't here.' We all knew that on board the aeroplanes we had been carrying lots of cash in US dollars, stored in metal attaché cases, and we also had cases of Glenfiddich whisky. I didn't know where all this had come from, but it had not been on the aeroplane when we left Stansted. So we knew we had access to tons of cash when it came to striking a deal with the commandant. Having got his attention, we announced: 'We are about to give you more money than you have ever seen in your life.'

This was Africa and these people were easily bribed. He immediately said, with a big toothy smile: 'My friend!' Next thing we knew, he was driving us towards the airport. We passed the famous temple in Luxor and he pointed it out like a tour guide as we cruised by. At the airport we were reunited with the other chaps, including Jock. A fistful of dollars appeared and was handed over to our captor. The ransom found its way into his pocket alone – one of the perks of being boss.

Jock instructed: 'Go and check the oil on our aeroplane.' In all the excitement, that was the first time I registered the heat of the day – it had soared to about fifty degrees. It was the first time in my life that I fanned myself and found myself getting hotter. It was a real shock, as if someone had slapped me in the face. Those DC-3 engines took gallons of oil. I had to climb up on the wing and it was unbelievably uncomfortable - burning hot. I checked the oil, put some in, and did the pre-flight while I was there. All the controls had locks on to stop the ailerons and elevators flapping around, and these I removed. I took off all the pitot static covers as well.

Then Jock and I climbed aboard, me in the left-hand seat. This aeroplane was so old it had a ribbon Gyro Direction Indicator (DI), rather than a normal round DI compass. This is unlike the usual more modern DI. If you want to achieve an increase of, say, 10 degrees from 190 degrees to 200 degrees, you turn right. But with the ribbon DI, you turn away from the indication. The compass rose is the opposite away round. It was a giro DI, but, rather than have a face like a clock where you could see all the points of the compass, the giro was on its side. The dial was like a ribbon, so you would just see your heading and 10 or 20 degrees either side. And you always had to remember to turn the opposite way. It was black with white writing, dead old-fashioned, wartime stuff really. I had never flown with one. The giro would spin up and you would have to make sure it was uncaged and therefore operational. It was a primary instrument in this old aeroplane. Anyway, much to my relief, we finally got on to the runway, again in formation, and Jock was going to let me take off with him covering me with the controls.

I wasn't at all intimidated at the prospect of taking off in the DC-3, not with Jock there. We would be flying with one call sign, which would indicate that there was a group of aeroplanes. With more than one call sign, it would have been chaos on the radio. Again, the plan was for Jock and me to lead and the rest would follow in loose formation. Our next destination had been revealed as Cyprus.

Even at this stage, I knew that Cyprus would not be in the direction of Arnhem, which is in the Netherlands. Was this still our destination? It was still a possibility, but as I taxied this old aeroplane on to the runway, I figured that we would be nowhere near where I expected us to be. We were ready to roll when suddenly there was a surprise command over the radio – 'You cannot get airborne, you have to come back to the terminal immediately.' Jock immediately sensed another conspiracy and commanded me not to return to the gates. He said: 'Bollocks to that, we are rolling.' Then the threat level was ratcheted up another notch. The local controller barked: 'If you get airborne, you will be intercepted and shot down.' A voice from one of the other aeroplanes intervened: 'I don't think this is a very good idea, Jock.' Jock was all for carrying on, but sense prevailed and we all taxied back in. As Jock had anticipated, this was simply another chapter in the extortion racket. They did indeed want more money. Obviously, one of them had said: 'I got a bunch of dollars out of them.' So more money was given out to others and we beat a hasty retreat before the whole of Luxor joined the fun.

We were heading up towards the coast and at some point we needed to turn right. We were in a successful tight formation and the American was out on my left wing. We all turned right, according to the plan, but the American simply carried straight on, prompting the comment 'Dozy bugger' from at least one cockpit. The American was the chap who kept saying he was a B707 captain, but every time I saw him, I was less and less impressed.

We eventually found him and the five of us regrouped for the landing into Larnaka. We were getting radar vectored and I was at the controls. We were told to turn to a heading, and I turned the aircraft accordingly, reading off this ribbon DI. Jock was alarmed and said: 'You're going the wrong way!' Immediately, I thought he must be right and started turning the other way. Then I realised I was correct - and Jock agreed: 'Oh shit, so you were.' While this confusion was being unravelled, the formation was following us and we charged around the Mediterranean in a big S-shape in the sky. Navigating with that bloody thing was a nightmare. And, of course, the state of our aircraft was pretty crap. There was even a big chunk missing from one of the elevators. In fact, all thc acroplanes were being nursed along, and overnight in Cyprus an engineer was told to fix what bits he could. All the aircraft were old dogs and they hadn't benefited from sitting idle in the African heat for five years.

We stayed at the Nicosia Hilton, where I had a huge suite, and discovered it had a lovely roof restaurant. All these pilots were much older than me, and they had wonderful aviation stories from around the world. The chap from South America, who had his own airline of DC-3s, turned out to have about 12,000 flying hours, just on Dakotas. I can shut my eyes now and still see them all around this big table under the Cyprus night sky, eating a lavish meal – no doubt all down to expenses. I remember drinking liqueurs of different colours and sitting there in awe, thinking: 'All these people are men of the world – and before tonight I've never even had a liqueur in my life.'

Later that night, we had another meeting, at which Jock announced: 'We're not going to Arnhem. We are not doing that. We are going to Israel.' I was a bit crestfallen. 'Well, that's me and my Sean Connery fantasy done and dusted,' although, I'm pretty sure, I didn't say it out loud. Jock went on: 'Since we left Uganda, there has been a raid by the Israelis on a hijacked aeroplane at Entebbe.' An Airbus had been hijacked into there, and Israeli commandos conducted an armed raid, called Operation Thunderbolt to rescue the passengers and crew, much of it seen by TV cameras. In a book subsequently published about the whole episode, called The Raid on Entebbe, it said five Dakotas were destroyed in the raid. Well, we were all witnesses to the fact that they weren't blown up at all. We had flown the five Dakotas out the day before.

So we were now, apparently, going to Tel Aviv. Once airborne the next day, Jock instructed everyone: 'We're going to keep tight and we are going low-level.' In next to no time, we were on the deck. It was like a bomber raid. I thought: 'What on earth is going on?' Then one of the aircraft at the back of the formation called up in a panic: 'Fast jet, low, three o'clock, coming at us!' Jock immediately ordered: 'Tighten up, tighten up.' My only reaction was: 'What's going on? This is amazing.' Out of nowhere, this big Phantom military jet screamed by in front of us, completely side-on, and banked away. I just had a chance to spot a Star of David on its side. It pulled up and circled above us and then it was joined by two or three other Phantoms, which had arrived with more stealth than the first. They were there to accompany us safely into Israel. We had been expected. We got ready to land and it was chaos yet again, with our aeroplanes overshooting and making a mess.

Jock taxied the aeroplane to a deserted spot on the other side of the airport as directed, nowhere near anything important. We shut down, Jock still in the flight deck, and I went back down the sloping fuselage towards the tail wheel and opened the rear door. Within seconds of the door opening, it started to become clear to me what the hell was going on. No sooner had I got the little set of three steps clipped to the side to disembark than a military-looking chap arrived, looked up to me, and said: 'It's nice to have them back.' I thought: 'Nice to have them back? So, all along, these were Israeli aircraft we have just flown through Egypt.' Egypt was a country seriously at loggerheads with Israel at the time.

Over the next few hours, the story unfolded. It turned out that at one point in the early 1900s, there had been talk of making Uganda the homeland for the Jewish nation by giving over a vast area of British East Africa. It was one of the options put on the table as an alternative to Palestine, but the Jewish people would have none of it, so the idea was dumped. After they settled into their new land of Israel, they kept good relationships with Uganda and even lent that country the five Dakota DC-3 aircraft for use by military paratroopers. At the end of this rather strange tale of inter-country co-operation, this operation was simply us bringing them back to their home base. But if we were expecting a heroes' welcome in Israel, we were soon disappointed. By now, the raid on Entebbe had taken place and it quickly became clear that our presence in the country was something of an embarrassment. So much so that we were all arrested and put under house arrest in Tel Aviv. Even at that point, Jock wasn't dismayed. 'Well, that's it, we've done the job,' he said breezily. 'We've got the aircraft here.' Clearly, Sean Connery was going to have to wait…

I had been to Israel before, of course. An idea crossed my mind that once we had extracted ourselves from this rather bizarre stand-off with the Israelis, I would part company with my flying buddies and spend a week or so with my friends in the kibbutz. By now, we all had Arabic stamps in our passports, and this, along with other reasons which were not made entirely clear, meant we were not welcome once the aeroplanes had been gratefully received. That night, we were told we were going to be put on the first scheduled flight out of Israel the next day, to wherever that flight happened to be going, and we could only hope it would end up being roughly in the direction of the UK. So we stayed the night in a hotel and were prevented from going out. I don't know to this day what was really behind this sudden change of Israeli sentiment, but one thing was certain: with the aeroplanes safely back in their hands, they just wanted the whole thing to go away. No one ever explained. It was just: 'Stay in your hotel.'

The next day, we were driven back to the airport. The first flight out was a Swissair DC- 10, going to Zurich, and, like it or not, we were going to be on it. As we were ushered out of the car at the bottom of the stairs, a rather tough military individual approached our little group and announced menacingly: 'Don't come back!' Once on board the aircraft, there was a palpable sense of relief, and soon the champagne was flowing. Our job had been completed, and done well. I hadn't a clue what was really happening, but one thing I did know: my pockets weren't exactly bulging with reward. In fact, I hadn't received a penny for my troubles since we left the UK.

I consoled myself that at least we were on a Swissair aeroplane, heading back to civilisation. Party over, we landed in Zurich and there we all split up. There was no great farewell. I never did know if we had been carrying arms, or anything else, into or out of Entebbe, and there is no way I would have been told if we were. After all, I was twenty-one, and my only real interest in the whole world was flying aeroplanes. I heard later that the mysterious chaps from the Swiss arms company who had gone with us to Africa had stayed on in Entebbe. Jock's advice was always: 'Don't speak to them, don't go near them.' And, of course, the whole film thing and dreams of James Bond drinking sessions turned out to be just part of the cover story. Yes, the film A Bridge Too Far was made that year, but did the Israelis sell the five Dakotas on to the film people – who knows? It's certainly true that in the film they did use military Dakotas, and ones with little holes in the windows. I saw the movie and always wondered: 'Have I flown one of them?'

Jock and I flew on to London, and by the time we arrived at Heathrow, I was already thinking: 'What's next?' Jock wasn't coming to Glasgow. He dipped into his pocket and said 'Here's your ticket with BEA', and he handed me a bundle of notes which added up to exactly £100 cash. That marked the moment I first earned any money from flying an aeroplane. I was chuffed to bits. As usual, he had his cigar hanging out the side of his mouth. I looked at him with renewed affection. He was just a wee guy and a terrible man with the women, a real hooligan. He handed me the cash and the ticket outside the Terminal 2 building, thanked me and said: 'Goodbye.' Then he turned on his heels and was gone. That was it. I never saw him again.

Absent friends

I flew back to Glasgow and by then I was not too far short of my required hours. I had been logging them throughout this escapade, although not as P1 because it was illegal to log DC-3 hours. Meanwhile, my mentor Jim Beaton, from the Glasgow Flying Club, had been going through a difficult time. He owned some of the club's aeroplanes and he had a bit of bad luck when a few of them were damaged in different incidents. He had disappeared from the club. My old pal John Frazer, the chap I had joined in the dodgy car business, was now driving a taxi in Glasgow. He was a bit of a wide boy, but I always found time for him because he shared my passion for flying.

One day, he said: 'You don't have that many hours still to do – why don't we hire an aeroplane dry?' This meant just hiring the aeroplane and putting our own fuel in it. 'We'll do a trip, and that will get you up to your 150 hours.' He found this Cessna 150 at Elstree which was so old it had mechanical flaps. All modern Cessnas have electric flaps, but this was the only one I'd ever come across without them. Soon after, I started to realise that John was a bit like my old flying chum Jock, in that he would get into trouble wherever he went. A few of the chaps at the flying club had warned me about him. Heeding not a word of this, within a week or two of his proposal we were on our way to Elstree to check out possibilities for my next adventure. There weren't many aeroplanes in Scotland for hire. In the south there were more and they were much cheaper.

We met up with this chap, who was keen on doing a deal with his cheap old banger of an aeroplane, and because there were only two seats, he checked us out separately. He took me up first, then he took John up and came back and said: 'Okay, that's fine.' Little knowing the risks involved, we were off to France, taking off from Elstree and heading south. Both of us were aspiring instructors, and the plan was that we'd fly all the way down to the south of France. When we crossed the border from the UK to France, we had to file a flight plan, but once we were in France that was no longer required. We flew on and on until we came across a few big thunderstorms near an airfield called Toulon Hyères, just north of the Mediterranean coast. We were getting chased by this huge thunderstorm, so we landed at the first airfield we could locate. When we landed, they were very unimpressed because it turned out to be a military airfield.

Arriving unannounced was bound to cause trouble, but we thought: 'Oh bollocks – we land, otherwise we're dead. Shout as much as you like!' Frankly, we thought we were going to die in that huge storm. To the people at the French military airfield, we were just a couple of inept pilots stumbling around France in this old Cessna. Anyway, they let us stay until the weather improved and then we nipped over to Cuers-Pierrefeu, where we slept in the aeroplane.

The next morning we found ourselves some breakfast in a little café off the airfield. Here my two per cent French exam result showed up when I unwisely attempted to ask for breakfast. I requested 'jam' which this French person heard as 'jambon'. I said 'Oui', and I got a large slice of ham. A little underfed, we flew the aeroplane on to Nice, just along the coast heading east. They gave us permission to land, even though it is a big international airport.

We hired a car and this prompted me to suggest to John: 'I'd like to go back to San Remo where I used to go as a child.' So we headed off across the Italian border to San Remo. There is a well-known casino in San Remo, which, of course, as a child I knew nothing about. John, ever the adventurer, said: 'Let's go in there.' So we went in and were immediately picked up by two professional girls. We were hauled on to the dance floor and while we were dancing, a bottle of champagne arrived. It struck me straight away that, first, it wasn't full and, secondly, we had the word 'mugs' written all over us. I dragged John off to the loo and told him: 'We've got to bail out.' So we did.

To top off a far from perfect evening, we then crashed the car. It was at night and, being Brits, we were just a bit vulnerable driving on the wrong side of the road in the middle of nowhere. We pulled out on to the wrong side of the road and into a head-on crash. It wasn't the most terrible accident, but the car was finished and the police arrived. The first thing they said was: 'Usually, by the time the British get this far down in France, they have learned how to do it right.' We said 'Well,

we flew down and came in the back door', and they thought that quite amusing. We weren't hurt and neither were the French people in the other car. It was just a good old crunch.
From Nice, we began the flight home. At one point, we stopped at a wonderful little airfield, typically French but similar to the one I had stopped at going up to Sweden. French regions were very proud of their local airfields, which tended to have one chap who did air traffic control, and he'd also drive the fire engine. They were immaculate, with a little terminal building, quite often a restaurant, and they were known for their local produce – so there would always be a fine meal on the menu. This airfield was just where we needed it to be, so we flew in and were made very welcome. The Cessna's range was only marginally better than our bladders. We stayed the night at the airfield and they let us roll out our sleeping bags and sleep in the open air, under the wing. It was dead cheap and everything was so easy. Quite a contrast to an airfield like Cranfield, where if you don't have your yellow jacket on, a sniper shoots you from some vantage point.
A French airfield of similar size to Cranfield is Le Touquet, a favourite haunt for British pilots because it's only an hour's flying time from the London area, and, more importantly, it's a friendly place to stop. The old Bristol Freighters used to fly in there from Lydd, operating the car ferry service. In the 1950s, they'd load up two or three cars and their passengers and fly them over. It is always a favourite destination of mine for a day out.
John and I had to clear customs and file a flight plan before we crossed the international border back to the UK. This we did in Dinard, and on arrival in Jersey we didn't get locked up, so we must have had everything in order. We got back to Elstree with the aeroplane still in one piece. We hadn't broken it, and about 15 or 20 useful hours had been added to my log. That got me up to my required hours because I had been logging at least half of our total flying hours. And the whole adventure hadn't cost us very much. Fuel was cheaper in France, landing fees were cheaper, everything was cheaper. We handed back the aeroplane and paid the owner in cash. He was happy – he had his aeroplane back without any dents. He knew that a lot of damage could be caused to an aeroplane with the thing still looking okay, so he completed a thorough inspection, and then off we went on our merry way.
Now, at last, I had my hours. My next challenge was to find the money for my instructor's course, in the region of £1,000 – a considerable sum in those days. In Scotland, there were not many opportunities to do an instructor's course. There wasn't anywhere in Glasgow, but that didn't

stop people at the flying club chipping in generously to help me cover my costs. My fund was further boosted by the £100 I had won with a rather unusual bet that I could get my weight down to twelve stone within an agreed period. The bookie friend who took the bet paid up when I got to twelve stone and said: 'That was such a lot of weight to lose – I'll give you 100-1, because you've done so well.' I had given him one pound and turned it into £100.

I decided Oxford was the best centre for my flying instructor's rating, so I applied to CSE, an Oxford training school which trained for professional licences. It was at Oxford that I met Bill Beadle, who was to become such an important person in my career development. The airfield was at Kidlington, near Oxford, and I stayed there for the duration as they had dormitories, mainly for the overseas students who would arrive for eighteen months to train as airline pilots. The flying school was split into two sections, one for the professional pilots' courses, and the other one I was attending – for private pilot and instructor courses.

I had to do a written test first and I thought: 'This will not go well.' Nevertheless, I got hold of the required books, and during the summer break sat at home in the dining room studying. In a few weeks I would have to go down to London to do the test, there being no other facility where this could be done. As I started to read, I had this sudden revelation: 'This is really interesting!' For the first time in my life, I had a huge feeling of 'I get this, I understand this. It's all interesting, and it's relevant and it's going to be incredibly useful'.

I had to achieve a 70 per cent pass mark in the multiple-choice test, so I studied day after day, reading through meteorology and all the required subjects. Then it was off to London, and I passed with 78 per cent. I hadn't even scraped through – I had passed! This was my very first experience of exam success. For the first time, I acknowledged that I could learn through study, not just through practical experience.

The new chief flying instructor at Glasgow was a first officer with British Airways, and a very different person to Jim. One day I had decided that he might be able to offer some guidance on my prospects of qualifying to become a flying instructor. I asked him one day: 'Is it difficult?' He replied: 'Oh yes. The flying instructor's course is very demanding, you have to do this, this and this.' Not convinced, I asked another chap, and he only confirmed my worst fears: 'Yes, it's very difficult.' But by the time I started getting into the prep for the London test, I realised: 'I can learn this.' I was finding it very interesting. I was going to be at Oxford to learn so many new things about my passion in life. It dawned on me that I actually knew very little. Because I had been flying for so long, I knew how aeroplanes worked, but I couldn't have told anyone much about induced drag, profile drag, and all the technical subjects I was reading about.

Bill, the chief flying instructor, had flown Mosquitos during the war, he had flown Hornets, and had written books on aviation, including some for the private pilots' course. And here I was reintroduced to Birch and Bramson, the famous Volumes One, Two, Three and Four. Those were the books I had read during my pathetic attempts at aerobatics over Loch Lomond some years before. There was another series of books, which Bill had co-authored with a chap called Dave Davico, another very senior examiner. Bill's examining number was X7, and he sat on the panel of examiners, so he was well and truly at the top of his profession. Dave, who eventually did one of my later flight tests, was also a respected author and examiner. Sadly, he was killed not long after my test, in a Cherokee in Carlisle where he was based. He was flying with a student, they went into a spin and they didn't come out. It was presumed the student had control, and whether he jammed his foot or had been pushing the wrong rudder, no one could establish. Over the years, I came to realise one very harsh fact: in a situation like the one Dave was in, one of the first things you have to do with a control problem like that is to give your fellow pilot a serious punch in the face – so he'll immediately stop doing what he's doing wrong.

The course was formal but very friendly. Bill would start with a briefing on, say, effective controls, and after his briefing I would have to present the same briefing back to him, in the same formal manner, using a whiteboard. It was back to basics stuff to start with, but it was a double lesson – this was exactly what I would be teaching my own students in time. I'd never seen it taught this way before. Jim Beaton would simply say: 'Jump in the aeroplane and we'll go and do effective controls.' No one ever stood by a board, as Bill insisted on doing. By the time I became an instructor, most people didn't do it Bill's way, and even today, sadly, that's still the case.

Push and pull

I didn't learn of Dave Davico's death until after I qualified as an instructor. Lesley Butler, a young woman from Glasgow who had been one of my instructors, was the first person I knew well who had died in an aeroplane when she crashed into a loch. When I examined closely what had happened, I became aware that her judgment had not been good and she had pushed the aeroplane beyond its capability. It's a bit like driving a car fast on an icy or wet road, when you have not yet reached a point of being completely out of control. There is that sort of grey area in aviation where I have been a few times. It effectively means it is in the lap of the gods whether you kill yourself or just damage the aeroplane.

In commercial aviation today, you never venture anywhere near that area, but sometimes you are still pushing a boundary and asking yourself: 'Should I be making this approach?' In 2014, on a passenger jet, I was shooting an approach at night into Zakinthos, an island to the south of Kefalonia, where there is a VOR approach. The runway runs almost north-south between hills on the west and a peninsula on the east. It was summer, and the weather forecast around the Greek islands included isolated thunderstorms. As we got near our destination, there was a tremendous storm to the north of the island. Greek air traffic is pretty poor, and in that vicinity there was no radar control.

I could see that the way air traffic wanted me to go would take me too close to the storm. So I told them I was turning left, to go towards Kefalonia to see if I could make my approach from that angle. They just said: 'Yes.' I worked my way around the island so I could come in from the south, and that way I could see the runway and the surrounding airfield. The thunderstorm was north-north-west of the airfield and flashing continuously. This was one of those really big active storms, and I was now speaking to Zakinthos air traffic directly. They advised me it was raining. I was more worried at that point about getting struck by lightning. I was worried too about turbulence, and I was concerned about a possible flooded runway and poor braking actions.

I had made up my own bit of airspace and constructed a hold 15 miles to the south. Then a German Hapag-Lloyd B737 arrived on the scene. I spoke to him directly and said this was where I was, and he, quite sensibly, entered a hold just above me. So now the two of us were going round and round in the holding pattern; he was 1,000 feet above me. I needed to decide: 'Do I shoot an approach or do we just divert to Athens?' If I went to Athens, we would be stuck there for the night. It was an option, but the passengers wouldn't even be on the island, and we would have to fly them back after a suitable rest period, so the pain would just go on and on. After a while, air traffic said the rain had stopped and the runway was dry, but the wind had changed. There was now a tailwind of about 10 knots, within limits, but not ideal in the conditions. I still had some concerns. I thought: 'The runway cannot be dry. It might not be flooded, but it can't be dry.' I knew I would be landing with a high ground speed, and once I had selected reverse I would be committed, because I couldn't then go around.

Years ago, a B737 landed, selected reverse thrust, and then the crew aborted the landing, cancelled reverse and went back into forward thrust to get airborne again. One of the engines remained in reverse and the aircraft rolled over and crashed. So the lesson has always been that after you have selected reverse thrust, whatever disaster is going to happen, it's going to happen on the ground.

My new plan was to now land on the southerly runway and land with a headwind. I was going to fly out for about 10 miles on a teardrop of about 30 degrees and come back round descending. So I descended to shoot the approach. I was overhead at about 2,000 feet, and, lo and behold, it was raining. My immediate reaction was: 'I'm over the airfield and it's raining, so it must be raining on the ground.' As I was heading outbound on the procedure, I was in and out of cloud, and there was flashing all around – we had found another storm on the radar and I was aiming for it.

Radar was showing red and I thought: 'I can't do this.' It was at this point that I was pushing it a bit. I turned early, which was fine because I was above the procedure, so I knew I was at a safe altitude. But by turning early, I had lost track miles to touchdown and was now high on the approach. I had my igniters on because I didn't want the engines to stop with all this water going down them. There was no risk of icing because it was summer, but there was a lot of water going into the engines. By now I had taken power off so we could get down and I said to the first officer: 'We're high, but I am happy continuing. If we can't get the height off, we'll throw it away and we'll go to Athens.' As I came round the corner, I asked traffic control again: 'Conditions?' Their reply: 'It's not raining.' So I was still high but coming on to the glide path, and was stable by 500 feet, and we landed - in an absolute deluge! At this point, the wipers were on full speed to clear the screen.

I had learned something while going into Ujung Pandang in Indonesia in a heavy thunderstorm: if you get to 50 feet and you can't see the runway, it's flooded, so go around. However, if you can see any of the texture of the runway, or the lights, there is just a lot of standing water, and it should be okay to land. The conditions in Zakinthos were far worse. We landed and I could feel the autobrakes cycling on and off, and they were struggling a bit. I had full reverse on, and eventually we stopped. It was teeming down. My immediate reaction after shutdown was just to sit there on the flight deck for a few minutes with the first officer. I whispered to him: 'Let's just have a little quiet time.' There were lightning bolts all around. The Hapag-Lloyd aeroplane had followed me round, and he got in as well. When his aircraft landed, it looked more like a speedboat than an aeroplane. So the question arose then and will no doubt arise again in the future: would I do it again? My considered reply to that was: 'Given the same conditions, I no doubt would.'

In situations like these, you are starting to get outside your comfort zone. When I was going into Ujung Pandang, I was flying a B767 carrying pilgrims back from Mecca and was heavy with a lot of fuel because we were sub-chartered to Garuda Airlines. They wanted us to get in on schedule, because if we didn't we were off to Bali, which was miles away, and the schedule would go to pot. They were happy for us to hold, in the hope we would eventually get in. Now, if you had stopped the aeroplane at fifty feet over the runway threshold and asked me 'Are you happy?', I'd have said: 'No, not really.' We were pushing it. I landed and again went into full reverse and we did stop absolutely fine. But in Ujung Pandang, every season, there are often local passenger aircraft abandoned down the side of the runway. This is another example of that grey area where you are using your ability, your experience. But, of course, early on in your career you don't have that.

In Dave Davico's case, his student was doing spinning. These days you don't spin any more for any licence, and most modern aircraft don't spin well anyway. Something like a Chipmunk spins but a Cherokee just goes into a spiral dive. With some determination, it is possible to spin a Cherokee, and at one time you were required to do it for a PPL. Then the syllabus changed to spin avoidance as too many accidents were being caused by spin training going wrong.

Effectively, every aeroplane has only one wing, but we do refer to the left wing and a right wing. Neither wing is perfectly matched with the other. There is always one that is more perfect than the other in any aeroplane. So as you get into a stall, one wing will stall first momentarily, and that wing will drop. In an older aeroplane, the chances are it will just drop into a spin, with very slow forward speed but great vertical speed. You can usually read what has happened by examining the scene of the crash. Was there forward speed or was it all vertical? You'd be able to tell by the shape of the hole. To get the Cherokee into a spin, you have to get to the stall, and give it a good heave back on the control to get it really stalled. Then you kick on full rudder. The aeroplane will tell you: 'You really are trying to stall me!' But you have to be quite aggressive. Maybe one out of three times an instructor would have to do it again because they wouldn't have kicked it hard enough. It's possible that Dave Davico was teaching a potential flying instructor.

When I was at Oxford, most of the students on the commercial pilot course were foreign and had the money to pay for their licence. Very few people in this country could shell out well over £100,000 in today's money to become a pilot. It was not unknown for very rich young men to come to the UK and play at becoming a pilot. They would live the high life in Oxford and spend the weekends in London – and just keep failing their exams. The training school didn't mind as long as the money kept rolling in. Thankfully – hopefully – they would never be let near an aeroplane.

In those days, all professional pilots started at PPL level on a single-engine aeroplane, then moved on to twin-engines and the professional syllabus. That requirement – to start on little aircraft – has stopped. You can now be a professional pilot without ever attaining a PPL. Your licence doesn't include the authority to fly light aircraft. You can go fly your jet but you can't fly a single-seater. Consequently, we now have airline pilots who have never been upside-down and never once scared themselves shitless in an unregulated environment. I believe they have missed out on some very basic, but very valuable, lessons.

The introduction to understanding aeronautics and everything that goes with it starts with good instructor training. When I was training, Bill would say: 'Right, we're going to do a briefing today on engines or constant speed units.' To start with, the constant speed unit propeller was a complete mystery, but by the time I had finished my course, I could confidently do a briefing on it. In an exam, I was asked 'Why would an aircraft have a constant speed unit?' I answered: 'It absorbs the power.' And: 'Why would a propeller have five blades rather than four?' I said: 'You either make the blades bigger, but then you would have a problem keeping the tip speed below Mach One, so to absorb all that power you need five blades.' He said: 'Yes, and another reason?' I said: 'I don't know.' He said: 'What about tip clearance (having five blades would mean they would be shorter and further from the ground)?' I said: 'Have you seen a Shorts SD-330? You can walk underneath the prop.' He said: 'That's fine – you've passed.'

I almost tripped up at the end, all because I hadn't mentioned tip clearance. The subject would have been relevant in a little aeroplane like the Mooney that I flew. It had a good-sized engine but it had a two-blade prop so you had to be aware of the tip when moving to and from grass and tarmac taxiways. If you porpoised on landing and didn't get the power on, you were going to take out the prop. On the aeroplane I have a share in, there is a three-blade prop, so it's much smaller with greater ground clearance – although one of our members did manage to damage it, porpoising on landing. As always, an aeroplane will bite, given half a chance.

All of this was the theory I was learning. I wanted to be able to confidently pass on this knowledge to my future students.

Bill briefed me and then I would brief him, and we would go up and fly the exercise. It might have been stalling, so he would start off in the classroom by showing the aerodynamics of the stall and the breakdown of the lift at high angles of attack. There would be briefings that weren't directly linked to an air exercise: engines or meteorology, for example. After every flight, we'd come back and have a debrief. During the first course, the AFI (assistant flying instructor) course, I lived in digs at Kidlington Airport, but when I returned a year later for my upgrade to QFI, Bill said: 'Come and stay with us.' The AFI rating allowed me to teach all the PPL ground and flying syllabus, but when I was airborne instructing there had to be a QFI (qualified flying instructor) on the ground, or, if he or she was airborne, in the local area. As an AFI, I couldn't send a student on their first solo, so my second visit to Oxford was my upgrade course to QFI.

The AFI course took me a month, March to April 1977 - part of the route I had plotted to become 'a pilot', which in my eyes meant a flying instructor. The course covered everything – how you taxied the aeroplane, how you got in it, how you left it. I thought: 'I didn't know any of this.' I learned even the simplest things, like how in a single-engined aircraft coming off grass on to tarmac on a taxiway, you meet the tarmac at 45 degrees. The theory behind this was that if you could get a main wheel and the nose wheel on to the taxiway together – should there be a dip where the grass meets the tarmac – there is less chance of the nose wheel sinking and the prop hitting the ground. But on a twin-engine aeroplane, you cross the taxiway edge straight ahead, at 90 degrees, so if the nose wheel sank down, the props, which are back on the wings, would be protected. It was wonderful tuition.

I did get my twin-engine rating eventually, although not at that point because I simply couldn't afford to fly a twin. A twin-engine Piper Seneca would cost nearly twice as much to hire. And I had no real opportunity to fly a twin, there being nothing in Glasgow other than a Piper Aztec, which Robbie Hunter flew on air taxi work. Scotland was in the middle of nowhere when it came to aviation. For my flying test, I was sent to East Midlands to a chap called Hector Taylor, who was equally as qualified as Bill. Bill's panel examiner's number was X7 and Hector's was X8, both ex-military ex-wartime pilots who had given their lives to aviation and to instructing.

At Oxford I had been flying a Fuji FA-200 aeroplane, which I had never come across before. It's a Japanese aeroplane and semi-aerobatic. I was famously not at all into aerobatics but Bill said to me: 'Take the Fuji over to East Midlands.' Bill, as my instructor, had already spoken to Hector and said: 'He'll be fine.' So the result of the test was pretty much a given unless I was really poor. Bill told me: 'I've spoken to Hector, and Hector likes a bit of aerobatics, so take the Fuji, you'll enjoy that.' I took the Fuji, flew into East Midlands, met Hector, and he said: 'Right, give me a briefing on…' He just chose a topic at random and I gave both a full briefing and a pre-flight briefing on that subject. Then he announced: 'Okay, let's go and fly it.' So I went up and instructed him, and suddenly he said: 'That's enough, young man. I have control.' We did some gentle aerobatics. I had to sit there, trying not to be sick. He flew back and landed it, and when he got out he simply said: 'Well done.' That was it, that was me passed, and I flew straight back to Oxford.

I had been signed off by Hector and was now officially an assistant flying instructor. As an AFI, I had to have a check ride on every aeroplane type I was going to instruct on. The same day I got back from my test, I flew with Bill and he did a test with me on a Cherokee. I hadn't flown with him in that before, and that was a type endorsement test, which I passed. That meant I could then leave as a qualified assistant flying instructor, able to teach on the Cherokee. Then came another landmark event – my first ever lesson. It was a thirty-minute lesson for take-offs and landings. My first student was a chap who owned a fabulous Indian restaurant in Paisley. From that day forward, every time I visited his restaurant I was treated like a lord. I was paid three pounds an hour - very little, but at least I was now earning real money. Students would make a booking, usually for an hour's lesson, and my name would be put in the instructor's column. I was the only full-time instructor so there was lots of flying to be had. By the end of that first month, I had done 40 hours, which was a lot when I was briefing and debriefing as well. Before this time, the flying club used to do what they called a 'hot turnaround'. This meant the prop was turning as the instructor climbed in to join his student, who had completed the pre-flight checks and got the engine running. Bill would have had kittens over this sort of training.

By now, I was living back at home and driving to the airport on a motorbike. Even at this stage, I had not a single jot of confidence in the idea that I would ever become a professional pilot. It just was not an option. What I wanted to achieve was my QFI rating. I wanted to be able to teach at night, I wanted to be able to teach the IMC (instrument meteorological conditions) rating, which allows you to fly in cloud. I wanted all the ratings I could get. And I wanted to do this as soon as I possibly could. I needed about 100 hours of instructing to do a QFI test.

Then, it came time to do my twin-rating. Nowadays, to do a twin-rating, you require 70 hours P1, an approved training course including flying and theoretical knowledge. To keep it valid, you have a proficiency check every year. Not back then, fortunately. You didn't actually have to do any course, but I was required to go down to Biggin Hill for a flight test. Once there, one of my first take-offs produced a surprise. When we got to about twenty or thirty knots accelerating on the runway, the chap I was with failed an engine without warning. As I was trying to control it, he said: 'No, you don't do that. If an engine fails on take-off and you are still on the ground, close the other engine and stop.'

A couple of days later, I went up for my test and passed. I don't know why I wanted to do my twin-rating but I was really pleased. Every year from then on, I had to do a certificate of test, and Group B was stamped by the Authority to say I could fly twins. My night rating had been chalked up fairly early on, after I passed my PPL. I could fly at night with passengers but I couldn't teach it. At that stage, I could only teach in the daytime, in good conditions. I couldn't instruct on twins but I could fly them. The summer came and off I went to Mull again - and came very close to killing myself. I was flying the Airtourer, an aeroplane which started out as the Victa Airtourer and then became the Glos-Airtourer. My pal Bill Wilson and I phoned David Howitt to say we were coming up. We said we were going to do a beat-up of the hotel… so there I was, an instructor, and Bill, a new PPL, and I was about to attack and give anyone watching a demonstration of low-level lunacy. I came in off the Sound of Mull and what I was looking at in front of me was the shoreline and a fence running along the airstrip boundary. The grass strip ran left to right and sloped down towards me. On the other side of the strip there were woods on either side of an opening where the hotel was situated. From our vantage point, we could see the impressive Swiss log hotel building very clearly, and its little garden, and then tall trees behind. My favourite much-practised trick was to come in from the sea with a view of mountains, hills and the glen as a big picture in front of me.

My plan, forgetting completely my new-found status as a flying instructor, was daring but well-rehearsed. It was to come in, lift over the fence, cross the airfield, aim at the downstairs bar, and then pull up and fly, at knife-edge, through the two tall trees. That I had done before. This time I thought, after whizzing between the trees, I would pull up to almost inverted (upside-down), and from that position downwind, pull through and land. A great plan and one I thought would look magnificent. Trouble was, no bastard should ever try to do something like that. The first bit went spiffingly well and I pulled up and, ever the showman, I noticed people were coming out of the hotel to watch. I ended up downwind with a lot of bank on. Although the aeroplane was just a two-seater, it was a powerful little thing with flaperons – they were very unusual, in that the ailerons and the flaps all moved together to produce an incredible roll. Anyway, I was upside-down-ish and of course I didn't really do aerobatics.

I was moving very quickly, way out of my comfort zone, and thinking: 'Okay, I'll just pull and we will drop down and land.' At that moment, I realised I had lost a lot of energy. I was just a few hundred feet up with no speed and I could feel a stall. Bill was getting anxious and saying: 'Scott, Scott..!' At that instant, all I could think about was Bill Beadle's voice saying: 'Whatever attitude you are in, in an aeroplane, the wing can be stalled. Whatever else you are doing, you have to un-stall the wing before you can do anything, otherwise you're flying a brick.' To this day I can remember being up there, looking almost up at the runway with the nose dropping and pushing fully forward and putting on full power. In the cockpit, the aeroplane had two levers: one was for the prop and one was for the power. So I pushed both forward and immediately heard the engine come up. The prop made a distinctive noise as it began to bite. The next thing I knew, we were sitting at the edge of the runway. We hardly rolled forward at all because we had no forward speed. We just stopped there.

I have no idea how we got out of that. It turned out to be my closest call. Even when I nearly looped into the sea at Greenock it wasn't as close as that. My ability to recover, by un-stalling the aircraft, was down to the fact that I was able to respond in a way which was the exact opposite of what you think you should be doing in that situation. Most people would have been pulling to try to get in, but that would just make the aircraft stall. We just sat there for a wee minute, not talking, and then taxied in.

The lesson I learned, which I later passed on to my students as often as I could, was that you must un-stall the wing no matter which way the aeroplane is pointing, or which way up it is. That was something the pilots of the French Airbus which hit the Atlantic in 2009 failed to do. They didn't un-stall the wing. It's fundamental. I believe that if I had been at the controls of that aeroplane, or if the captain had got back to the flight deck sooner, it wouldn't have hit the water. The first officer was pulling, and the indicators were showing it wasn't slowing down. I have flown a jet with no airspeed indicators with another captain, who spent the whole flight saying: 'What speed are we going?' I said: 'Forget the airspeed indicators, select the appropriate pitch and power to achieve our flight profile.' The French pilots had a couple of minutes before they hit the water and they were confused. They were being told by their instruments that they were flying slow and fast at the same time. If pulling back wasn't slowing down the high speed indication, they should have ignored that and pushed the stick forward to try to un-stall the wing.

You learn things about flying all the time, but my scary escape over Mull was mostly down to luck because I don't know how I landed it. I should have died that day, there's no doubt about it. The situation I found myself in was way out of my capability. Of course, if I had died that day, I would have taken Bill with me, and I still have a very bad feeling about that. Sitting slightly stunned on the runway immediately afterwards, I thought: 'Damn, that was not responsible in any way.' This was me wearing my 'arsing about' cap, not my 'instructing' cap. We taxied in. Dave Howitt, who flew model aeroplanes and was very knowledgeable about aerodynamics, had watched all of this, and he knew better than most of the onlookers what a lucky escape we had had. He had analysed the whole thing exactly. The other people watching from the hotel would have thought: 'That is magnificent.' When we finally stopped, David was standing there holding two whiskies, as if to say: 'You'll be needing these.' He knew. So away we went and got drunk…

Almost unbelievably, around this same time and again in the Victa Airtourer, I nearly totalled myself over the same bay on the Sound of Mull. This time I was alone and instead of aiming for the hotel, I was out over the bay, parallel with the airstrip. Whatever stunt I was trying to pull, it ended up with me heading at an extremely steep angle straight for the sea – at high speed and at an ever-decreasing altitude. I could feel the aeroplane starting to stall because I was pulling so hard. Suddenly, I thought: 'I can't get out of this!' I could see the surface of the water looming closer and closer. The feeling was similar to my first near-miss over Loch Lomond when I attempted aerobatics and thought I was going to die. Now, almost exactly the same thing was happening. I was pulling, but at the same time trying to release the backpressure on the control column to stop the onset of the stall and keep the wing flying. I just missed the water as I whizzed across the shoreline and climbed away over the village of Salen. After a few seconds, I smiled and thought: 'Phew, I'm alive!' That was the last time I ever put my life at risk like that.

Flight to the Finnish

In my role as an instructor, I was conducting quite a few check-outs of PPLs at Mull, as the weather conditions can make the already tricky airfield very demanding. My log book records one trip with friends in August 1977, flying around the island of Staffa. We went up and did spinning. No doubt I was showing off, because if you are a passenger in a spin you know that you are about to die. It is a scary and violent manoeuvre, and looks very, very impressive as you see the world spinning around.

I worked really hard that summer, flying both for fun and teaching. In October 1977, I went back to Oxford, and spent three days achieving my upgrading to full instructor. I knew that passing my test would give me the authority to send people solo. At that stage, I had nearly 500 hours P1. So now I had become a QFI. Back at work in Glasgow, I was getting fed-up with the chief flying instructor. I heard they needed a QFI at an airfield called Stapleford Tawney in Essex, south of Stansted and just inside the M25. Most usefully for me, the Underground went out to Theydon Bois, which is close by. Stapleford needed a second QFI to cover the flight training.

Ruth and her friend had said I could always stay in their flat, so I had a way of staying in London. Being able to fly from a little airfield, where the whole operation was much more laidback, was very appealing and Stapleford had two grass strips, which I first discovered when I went for an interview and test flight. The chief instructor interviewed me and took me up in a Condor, an aeroplane a bit like my old one I crashed, which was also wooden with a tail wheel and very basic. They taught people from the London police school at the airfield using the Condor. This meant they needed someone with tail wheel experience and a QFI, and that fortunately turned out to be me. He took me up on a twenty-minute flight and showed me the local area. The airspace was very much more congested in the south east. You had to keep a good look-out for other aircraft, and also not wander into another airfield's zone. You could see the River Thames quite clearly, but we flew out east, which they used as their local flying area, as we had used Loch Lomond. At the end of the short flight, he announced: 'You've got the job.'

Stapleford was a real flying club with lots of flying going on, and a busy restaurant and bar. The chap who owned the club, John Chicken, also owned the airfield . He lived in a bungalow in the corner of the airfield. There were hangars on the site which had obviously been there from the war, and two grass runways. The clubhouse looked a like an old military hut. The airfield, originally known as Essex Aerodrome, started operations in the early 1930s, and Amy Johnson flew for the resident airline there, Hillman Airways. Before the war, the RAF used the field as a flight training school, and its most famous student was J.E. 'Johnnie' Johnson, who became the RAF's second top-scoring pilot during the war. When the war started, the airfield became RAF Stapleford Tawney, and was a satellite station for RAF North Weald, one of the famous Battle of Britain fighter bases. A charter company was based at the airfield flying Aztecs and Navajos, which were twin-engined aeroplanes seating up to ten passengers. Eric Thurston owned the charter company, and he used to drive around in a big Rolls-Royce. It became quite a large charter company, called Thurston Aviation. Some of the chaps who later joined Britannia started with Eric, and he opened a school at the airfield where he would teach twin-ratings and instrument ratings.

In my day, if you happened to be a big shot East End gangster and you needed an airfield, this was probably the airfield of choice. There were always some very interesting characters about the place. People just piled into the aeroplanes without too much regard for the pre-flight briefing and preparation. When the weather turned bad, the whole place was a mud bath. I started briefing people properly with my instructor's hat on, but as time went by, my briefings became less formal, I am sad to say. There was one chap, an older man, who was great fun to be with called Fred Wells, known as Oily. He became the chief flying instructor. He and I spent many evenings drinking at the flying club's bar, where I would sometimes sleep. I moved into a room in a large house close to the airfield, but I still saw Ruth regularly.

Another chap, who became a pal of mine there was Chris Sales, who rose to become a captain with Britannia, like me. At Stapleford, one of our tasks would be to check out people going across to Le Touquet in France. Quite often, private pilots are a bit unsure over new territory, so they'll ask an instructor to go along, particularly over water. There was a lot of flying going on, even throughout December and into January when the weather was generally poor.

One day, I was with a chap who owned his own twin. I was not a twin instructor but I had a lot more experience than him. I was the commander but it didn't say anywhere that I had to sit in the left-hand seat. We were flying west, he was flying the aeroplane, and we were in cloud. This was legal because we were not in controlled airspace and I had an IMC rating. He was not very good, and I didn't have that much twin-engine experience. On top of that, there were a couple of other chaps in the back of this Apache, which was fairly underpowered. Suddenly, the pilot said: 'There's something wrong.' I looked at one of the rpm gauges and it was falling. Of course, it would fall if the engine stopped, but my immediate thought was that it shouldn't be dropping as much as it was, because the propeller would automatically change pitch. I told him: 'That engine's failing. It's seizing.' I took control and closed down the engine and feathered the prop. At this point, the owner was screaming in panic.

I told him to be quiet but he had completely lost it. The guy sitting behind him grabbed him and stopped him screaming. Now we were on one engine and we were still in cloud. There is a thing called the blue line, on the air speed indicator, and it indicates the best speed to fly on one engine. My twin-engine training was fairly minimal and I wasn't absolutely sure about this blue line speed. I might have been flying a little fast, but I couldn't maintain altitude, even with full power on the live engine. It should have been able to fly on one engine but at that moment it wasn't. I was still in cloud, heading west, so I put out a Mayday and began speaking to Bristol. They responded, telling me where to steer, which I was trying to do but I couldn't control the aeroplane. It was coming down and was breaking out of cloud, and I could see fields. I was now at about 800 feet - all this drama had started at about 2,000 feet.

The cloud was breaking up and I could see fields and I was beginning to think: 'Okay, we're fine. I'll just treat this like a Cherokee. I'll shut the other engine and pop it into the field, job done.' Bristol had given me a heading towards them, and I steered that heading, still descending. I was out of cloud, I could see Bristol on the hill and the approach lights were almost at my level. We landed on the runway and stumbled in. As I taxied to a halt, I thought: 'Have I overreacted? Was that engine seizing?' It was obviously failing, but I had chosen to close it down when it might still have provided some power. On examination, it was discovered that the oil sump drain plug on the right engine had come out – a fault which was 'impossible', seemingly. All the oil was pouring out of the engine and it had been seizing at the point I had shut it down.

Oily – Fred Wells – flew in from Stapleford with another aeroplane that same day. He filled up the failed engine with oil and asked if I would fly it back. I said: 'I'm damned if I'm going to fly it back.' So the four of us climbed back on board and Oily flew us back that night. Oily advised before we left: 'Get the guys to lean forward when we get airborne to get the C of G (centre of gravity) forward, because if the centre of gravity is too far back, it won't fly.'

Not long after that little drama, Oily organised a ferry flight for us to go to Helsinki to bring back some aeroplanes. 'Are you up for it, boys?' Chris Sales and I said yes, and another chap also wanted to go. One of the more colourful, gangster-like characters around the airfield was a chap called Big Harry, who I believe was a hitman. He looked exactly as you'd expect a hitman to look. Harry had learned to fly and had his PPL, and I was teaching him for his IMC rating. He always presented himself to me as a nice guy, and the only tangible evidence that this appearance might have been deceptive lay in the fact that he always had to be back at a certain time – to sign into the local police station. Big Harry came along for the ride to Helsinki. He wasn't flying the aeroplane, he was with Oily. So there was Oily and Big Harry, me and Chris and someone else. We were going to bring four aeroplanes back from Helsinki – although I never confessed that I didn't even know where Helsinki was. To start with, we all had to get up to Luton, where we flew out on a B737 with Britannia Airways. I knew very little about Britannia at that point, and it was the first time I had flown with the airline. I can remember seeing the captain coming out of the flight deck, and he walked down the aisle to the toilets, at the back, wearing his white cap.

We got into Helsinki and the weather was awful. This was the beginning of March. As soon as we arrived, there was all sorts of hassle. Oily was trying to get the aviation paperwork for the four aircraft straightened out. Then he announced: 'All right, I've sorted everything out.' I had a strange feeling that what he really meant was: 'Let's just steal the aeroplanes.' Back at the airport the next day, we found these aeroplanes were under heaps of snow. So we brushed them down and found there was a single-engine Beach, two Cherokees and a Grumman Yankee, a little two-seater aircraft – a horrible thing – which was assigned to me. We refuelled, fired them up and off we went. The first leg, Helsinki to Turku, was an hour and thirty-five minutes.

My aeroplane was OH-AYB, which was Finnish-registered. It was called an AA-1, a single-engine aeroplane. The fuel gauge was just a long, thin glass tube, down the side where my legs were, and as the fuel was burned, the level in the tube went down. It was the same idea as the fuel cork on my Jodel. We were flying in loose formation over the pack ice in the Baltic, when suddenly my engine started faltering and one of the other chaps on the radio informed me: 'You've got smoke coming out of your engine.' I didn't have a map with me, but someone else did, and they quickly advised: 'There's an island way to the north.' So we all turned for this island and landed there. That was the time when, ever the showman, I thought: 'If I'm going to hit the water and die, what shall I say over the radio? I can't just crash!' I thought: 'I'll just say 'splash!'' The rehearsal never came to performance, thankfully, and we all got to the island safely.

We were lucky enough to come across an engineer. The locals had seen us fluttering, like lost sheep, as we descended into this place with smoke coming out of my engine. The engineers found a couple of dead plugs, which was why my engine was faltering. Helpfully, they said an SAS DC-9 was due in that evening from the mainland, and they would organise for replacement plugs to be put on the flight. We made our way to the Savoy Hotel on the island and all went out for a drink. Oily proceeded to get drunk in double quick time.

I became aware that there was a lot going on in the room. Oily was busy eyeing up somebody's woman and the locals were getting noticeably cross. There wasn't a big crowd but there were enough upset locals to be concerned about as Oily zigzagged off to the toilet. Big Harry, ever the pro, was watching all of this. Just as a bunch of local guys headed for the toilets, thinking 'We'll have him now', Harry said to me: 'I'm just going in to look after him.' A few minutes later, Oily came out in one piece and Big Harry emerged too – but nobody else. Harry said calmly: 'We're going now.' Later that evening, Big Harry took me aside and said quietly: 'I won't be out for long when I get back to England because there's been a murder. I didn't do it but my name's on it, and they're going to have me.' He knew he was on borrowed time. 'It's either that,' he said sadly, 'or I'll be hit.'

Back in the hotel, the party continued in one of the rooms – by party, I mean we broke up the room and had the door off its hinges, completely destroyed it. Being honest lads, we got up at about four o'clock in the morning, got into the aeroplanes and headed off. Later that day, we were approaching Sweden and there were two huge aerials north of Stockholm. As we passed by, between the two aerials in cloud and in formation, we all thought it was lucky that we didn't fly into them. We all climbed another 1,000 feet. We broke out of cloud and Chris and I, for no particular reason, decided to do some top-level hooliganising. We were attacking anything that moved on the ground. We went at this chap in a fishing boat, and laughed as we saw him drop into the bottom of his boat as we passed over at 10 feet. Even people on top of dykes with their bikes were fair game. Even better, we had to climb to get over the dykes – we couldn't help ourselves. That's when we realised we were passing over a group of tanks and army vehicles and we were now in a danger area. We were attacking an army training ground in Denmark.

Fun time completed, we joined the others, heading for a small airfield near Schiphol to refuel. Chris and I had been using a lot of fuel messing about, and he called up and said: 'I'm running out of fuel.' At the time, we were flying down the coast and we couldn't do a forced landing on to the huge sand dunes inland from the beach. I was behind Chris, so I said: 'You go in on the beach and I'll come in behind you.' The other aircraft kept going and landed at the airport, as we prepared to go on to the beach. I put a Mayday out, and a British Airways aircraft acknowledged it on 121.5. I told him: 'There are two of us going in on to the beach.' We landed, and a short time later, we looked up and there were police coming from all directions. They probably thought they were about to pounce on a gang of smugglers. We were taken somewhere and told we couldn't leave. They didn't realise we were just a couple of real twerps.

The next day, they gave us some fuel and the authorities finally agreed to let us fly off the beach. But now our presence in the Netherlands was official, not at all what we or the other chaps had planned. The authorities wanted us to fly to Schiphol, the main airport in Amsterdam. The weather was very poor with rain and low cloud, but they wanted us off the beach. They insisted on putting a local instructor in with Chris. Back at the beach, we were still being ordered about, and they commanded me: 'You must not go until thirty seconds after the other plane gets airborne.' The weather was getting really foul and I thought: 'If I wait, I will lose sight of Chris.' As soon as Chris had moved forward about eight yards, I refused to wait any longer. I just called out 'Bye!' and followed him into the air. There was no problem getting off the beach but we were soon flying in cloud and I was finding it difficult to keep in close formation with him. If I lost him visually, I was in big trouble. I hadn't done much tight formation flying, and I was catching him up and then trying to slow it down by closing the throttle and pulling out the carb heat, to get as little power as possible. Then I would drop back and have to apply lots of power to keep up. It was all very crude, but effective. We both landed safely into Schiphol, the others met us there and we stayed the night at a local hotel.

Next day, the plan was to fly back to the UK. The other two aircraft departed their airfield, and Chris and I were going to take off from Schiphol. We lined up on the runway together for a formation take-off, but as we started the take-off roll, my engine again began to run rough. We both aborted the take-off. Not a little frustrated, we taxied back in, parked the rogue aeroplane, left the keys, both climbed into the healthy aeroplane, and departed with not a backward glance. We just left the stricken aeroplane behind and I thought: 'That's it – I quit!' We flew to Southend and returned to the UK, having lost one out of four aeroplanes.

When I got back home, I could reflect on another episode, full of fun and adventure. At least, it was fun right up to the moment we all received a letter from the CAA, the Civil Aviation Authority, declaring: 'We have had this report from the following authorities – Finnish, Dutch, Swedish – a litany of low-flying, aircraft left illegally, without this and this…' I thought: 'Damn, I may not have a career in aviation after all.' The other chaps didn't really care. Big Harry shouldn't even have been there. He wasn't allowed out of the country, so how the hell he managed to give the authorities the slip, I've no idea.

I came up with a plan to extricate us all from the clutches of officialdom. I said: 'Look guys, I will write to the CAA on behalf of us all.' I wrote a letter acknowledging as much as I could, apologising profusely and trying to explain away our crimes, without being completely open. I thought I had addressed enough of the issues to get us off the hook. I sent the letter off and soon had a reply, which threatened: 'Don't ever, ever, ever do it again. And it's on your record.' So that was a bollocking from the Authority, but, on reflection, I felt we got off surprisingly lightly under the circumstances.

Goodbye...

My partner in crime on the Scandinavian jolly, Chris Sales, was an aspiring pilot, as I was. So there came a point where, like me, he had to take the CAA warning seriously. On the other hand, Oily Wells, who was a QFI, had no career aspirations, so anything from the CAA caused him no concern whatsoever. The bollocking did have an impact on Chris and me. I had done quite a lot of instructing hours, about 500, and I was a QFI, so I was well on my way to being able to take up a position as chief flying instructor at a flying club somewhere. I had to keep my nose clean. The other instructors at Stapleford didn't really want to be flying instructors at all, they just needed the hours so they could pass their exams to join an airline.

When I went to Stapleford, all I wanted and expected to be was a flying instructor for the rest of my career. I suppose, as the months passed, I started to consider the possibility I might achieve my commercial pilot's licence, the CPL. I thought I would never be able to join an airline and fly something big, but I might be able to fly for an air taxi company: maybe a six-seater Aztec, or if I was really lucky, a 10-seat Navajo, an aircraft with a real door – an aircraft you didn't have to climb on the wing to get in. But I would still need a CPL and have an instrument rating to fly those aeroplanes. The CPL would entitle me to fly any small twin-engine aircraft, up to about 18 seats, in command. Above that, up to about 25 seats, you needed a senior commercial licence, and above about 25 seats you needed the ATPL (airline transport pilot's licence) – but that was all up in the clouds. With a CPL, I could fly any aircraft as a co-pilot.

Another benefit of being a CPL was that I could teach pilots to become flying instructors, like Bill Beadle had done for me, or instruct pilots for their instrument rating. I could do both working for Eric Thurston's charter company at Stapleford. They would do freight runs to Cologne at night, carrying mainly newspapers, and also carry out instrument rating training. Occasionally, Eric would come over and ask if I wanted to join the pilot on the flight to Germany that night, and I would sit in the right-hand seat. I had never come across this sort of operation, being operated by light twins, but it continues to this day. I still hear over Olney, at around one in the morning, the drone of some small aeroplane heading for East Midlands full of mail.

When I asked 'How hard is it to be a flying instructor?', people would tell me: 'Very hard.' But one by one, as I tried the many different tests, I usually passed okay. It took years to realise that if I worked really hard, I could achieve what I was aiming for. I have learnt that most people like to make things look as difficult as possible. When a new pilot asks me now about a career in aviation, I say they should take each hurdle one at a time. Don't look down the track – concentrate on the next jump.

I began to consider seriously my prospects of achieving a commercial licence and decided I would go to Oxford again, even though Bill Beadle wouldn't be involved in the ground training. I would be required to do nine written exams, including navigation, meteorology, instruments, Morse code and air law. The pass mark for most of them would be 70 per cent and once you passed any section of the syllabus, you had to complete all ground subjects and all the flying tests within a year. If you didn't, you had to start all over again.

On top of the ground subjects, there would be four flying tests, called GFTs (general flying tests). The instrument rating flying test was separate, and outside the requirement to complete within a year. For that, I would have to be flying a twin, which at that time was costing me a couple of hundred pounds an hour. I was going to need about £30,000 to pay for it all and I got a loan through the RBS bank. I started my CPL/IR course at CSE Oxford in the summer and booked in for the October exams. I passed everything, except air law and Morse code. They were not bad ones to fail as it turned out; I could re-sit them a month later. I failed Morse because of the lights element of the test. The audio was fine. For days before the test, I was driving round with tapes going 'da da dit dit da'.

In the test, the lights were done so slowly that it made it difficult, whereas the correct speed for Morse was easier to understand. There is no aviation relevance for Morse code using lights nowadays, but we use audio Morse to identify our ground navigational beacons today on every flight. I was learning things about Loran and Decca navigation systems, which were no longer used but were still in the exam syllabus. All these things were just another fiery hoop I needed to jump through, so jump I did.

I passed the exam about a month later and all my commercial writtens had been successfully completed. I still needed to do the flying tests, including a GFT 4, which is a night flight test. The CAA would allow certain instructors to do that because I don't think they fancied flying around themselves in a single-engined aeroplane at night. I did that at Stapleford and passed. I had to do a cross-country flight in the day and some in-flight manoeuvring, including stalling, but not spinning, steep turns and forced landings. So those were GFTs 1, 2 and 3 and 4. Then I went to Stansted where CAFU, the Civil Aviation Flight Unit, was based. I flew over there and did a test with a Captain Eastwood, better known as 'Chopper' Eastwood. I failed GFT3, which was my first failure in an aeroplane. GFT3 was a cross-country test. To start with, I did everything he expected me to do, and then it was time for him to put me under a hood, to ensure I couldn't see out of the aircraft. He flew me around for a while, and then took the hood off. I had to identify where I was.

This was in November, and the countryside below was covered in snow and all the landmarks were gone. It turned out I was over Biggleswade, although I would never have known. I had three goes at the damn thing before I cracked it in March the following year. The knack to passing 3 was good map-reading: the ability to look out of the window and find yourself on the map. There is a circle of probability, taking in how long you had been under the hood and how far the aircraft could have flown, but you still needed to find yourself on the map. We were flying over Suffolk, which is incredibly flat, and it had snowed, so I didn't find Biggleswade that first time. The name still grates on me to this day.

Before I finally got through all my flying tests, I found time to achieve another landmark in my life, which put all my flying achievements in the shade – I got married. Ruth and I were married in Swansea, in Clyne Chapel, on December 28th 1978. We had no money so we didn't want any presents - just cash from our parents, friends and relatives to go towards my flying. On the morning of my wedding, I got a bit tipsy. The wedding was at Clyne Chapel and friends and family had come down from Scotland. My pals gave me a few too many Drambuies before the ceremony, and I was feeling a bit woozy.

Afterwards, Ruth and I headed off in our ancient purple Austin 1300GT, suitably decked out by our guests, on our honeymoon of two nights in Bath, which was all we could afford. And that was us married.

I didn't have any spare money. I only had money for flying, and I was going through that pretty quickly because every re-test was costing me hundreds of pounds. Ruth had to buy her own engagement ring. She went off to Hatton Garden in London and came back with her 'Jammie Dodger', a ruby and diamond ring in a traditional setting, so-called because it reminded her of the jam and cream in the biscuits.

By the following February, all my exams were done and in March I finally passed GFT 3. At last I had a commercial pilot's licence - but a commercial licence was useless without an instrument rating. For that, I had to borrow money from Ruth towards the end of my course as I was running out. I borrowed £1,000 she had saved up to buy a car. I don't think I ever did pay her back, but I needed that money for my flying. Flying a twin was costing me something like £250 an hour and the cost of my training was mind-boggling. It would take me half an hour just to fly to the test, the test was an hour's flying time, and you had to pay for the test as well. I was haemorrhaging cash at this stage.

The GFTs were in a single-engined aeroplane so they weren't as expensive as flying a twin. The expensive part, the twin flying training, started in the New Year. Having started my training for my instrument rating, I was logging about 20 flying hours - at £250 an hour. Ruth used to come over sometimes to Elstree where the training was based, at the London School of Flying.

I used to go flying and she'd come up occasionally with me because the twin was more stable than a single-engined aeroplane, and she didn't feel quite as airsick. Quite surprisingly, I passed my written exams for my professional licence, and there were just the flying tests to come. I thought: 'I've always passed everything I've done flying, the flying will be fine.' The flying turned out to be far from fine. I went for my first attempt at instrument rating and failed. Again, I could fail a whole lot of sections of the test and still proceed, and I failed one, the ILS (instrument landing system), where I had to fly on one engine with one closed down. I was training at Elstree aerodrome, north west of London, and had to fly the aeroplane to Stansted to do the test. Usually, people would fail something like the hold or the NDB (non-directional beacon) approach, which is much more demanding, and I passed all that.

By now, I was starting to apply for jobs. That March there was quite a lot of work about for pilots and at one point I went for an interview in the Midlands for a large manufacturing company. They had their own Super King Air and Queen Air aeroplanes and they wanted a co-pilot. I went for the interview at their own private airfield. I thought the King

Air was a beautiful aeroplane - a 12-seater with twin-turboprops. It was not unusual for large corporations such as Shell, Rolls-Royce and JCB to run their own aeroplanes. I came away from the interview feeling really keen, but I didn't get my instrument rating quickly enough. Undeterred, I applied to other companies, in the UK and abroad. I spent hours typing away on an old typewriter and sent off dozens of letters.

The lack of money kept coming up again and again. I couldn't stop or postpone for a while because the deadline to complete the licence was always there. Even the poor flying conditions, which meant cancelled training flights or flying tests, caused stress. In the UK, I applied to Air Anglia, based in Norwich, BIA (British Island Airways), Dan Air, and Loganair in Glasgow. I didn't care which airline offered me a job, as long as I got one. BIA flew Handley Page Dart Heralds, Air Anglia flew Fokker F27s, Dan-Air flew Hawker Siddeley HS748s, and Loganair smaller aircraft, such as Islander, Trislander and Twin Otter. The first company to offer me an interview was Loganair. They all did in the end but I chose Loganair because they offered me £1,500 cash in hand to help with removal expenses. That money could keep us going and me flying. I was offered a job with Loganair as a second officer with one stripe. I was in Swansea, preparing for the re-sit of my instrument rating when I got a call from the chief pilot saying: 'Can you start Monday? We've changed our mind – we're putting you on the Shorts SD-330, not the Trislander.'

The SD-330 was a new type of aircraft. Loganair was going to be the first airline to operate it. On the first Monday of my new career, I was off to Belfast with a group of fellow Loganair pilots. The ground course would last a month and would be run by the manufacturer, Shorts. On the phone call I had asked: 'What about my instrument rating?' He said: 'We'll sort that out later.' So the good news was that they were going to start paying me straight away, and I had the £1,500 to join. The course was to be run at what is now George Best Belfast City Airport, but then it was owned by Short Brothers. Just before the Loganair offer, I was really struggling financially. We'd used up Ruth's car money, and I had sold my precious Second World War leather flying jacket. With my new job secured, Ruth and I went back to Glasgow. On the Monday, I headed to Northern Ireland for a month, and the plan was that we were going to come home at weekends. We were all in a small hotel, outside Belfast on the coast, which was probably deemed safer because of 'the Troubles' at that time.

Early on the course, I met a fellow new Loganair second officer who was to become a lifelong friend. His name was Camille Bibby, a British Arab of Lebanese origin. He was brought up in Beirut and had been flying as a steward with Kuwait Airways. He took the job of steward simply to get enough money to pay for his pilot's licence, which he did at Oxford. He had done the whole course there and then took a job based in Cardiff with a company called Air Wales, but within a month it had gone bust, so he joined Loganair. I had more hours than him because of my flying as a PPL and flying instructor, but he had worked for another airline which I thought was impressive. Cami had just married Karen when I met him, and they were renting a flat in Paisley, near Glasgow airport.

We were going to fly the very first one off the production line, after the test aircraft. Its registration was Golf Bravo Alpha November Alpha, but we weren't going to be introduced to the aircraft until we had finished the ground course, one month later. Being a manufacturer's course, they went into the aircraft systems in tremendous detail. It was quite mind-blowing how much detail they went into – a depth of knowledge that we don't even remotely get near today. When I did the B767-200 course in the mid-eighties, the book said the aircraft pressurises when the doors are closed and the engine thrust levers have advanced by so many degrees. These days all it would say is: 'After engine starts, the aircraft pressurises.' There is much less technical detail now during a type conversion course.

History has shown that very few aircraft have crashed due to insufficient pilot technical knowledge. When I did that Shorts course, and during my early jet courses, it was 'chalk and talk' by engineers, at the front of the classroom, going through diagrams. Now you sit in front of a computer and go through it yourself with an instructor in the background to answer questions when you get stuck.

I had a sinking feeling as the course progressed because the other chaps had all done other courses and a lot of them were ex-military, so the level of experience in that classroom was amazing. I asked an ex-military captain why he had joined Loganair and not one of the large jet airlines. He said: 'I don't really fancy flying a spam can to Palma every day – this sort of flying sounds much more fun.'

Because there were no Shorts aeroplanes yet certified, the CAA came to Loganair for us to sit the exam. The examiner came in and said: 'Right, this is the exam. You've all done jet engines, DC electrics, propellers?' I immediately said: 'No!' Somebody else said: 'No.' The examiner announced: 'Right, I'll tell you what, you guys carry on and I'll go and get the questions by phone.' He disappeared for about an hour and in that time we all conspired to cheat. Once we had all finished the paper, one of the chaps piped up and said: 'We can't all just pass with the same mark - we had better change some of the answers.' Some of them unwisely changed too many of the correct answers – and failed. I changed not a single one. I had also been studying for the general subjects, DC electrics, constant speed unit propellers and jet engines. I failed the propeller test and that is when I had to go to Redhill to re-sit. I'd been swotting up on all of these subjects while I was on the course and I could have written a book about DC electrics. Anyway, I passed them all.

After all of this, we still had not flown the new aeroplane, though we had seen it in construction towards the end of the course. There was huge time pressure to get the flying done and the test passed because the aircraft was to be flying on the line within a couple of weeks. During this time, all the Loganair captains had done their flying training and their final flying test on the aeroplane, but we three co-pilots hadn't because we were, after all, just co-pilots. The chief test pilot at Shorts was going to do our flying test. Jamie Bailey, the training captain at Loganair, took us to Belfast. Jamie had this rather endearing method of teaching, whereby he would use an officer's stick to hit us to emphasise a point, or to point out we had got it wrong. It was all done in fun but it could sting a little. Alan Stanford, the other co-pilot, flew from Prestwick to Belfast. Frankly, the amount of training was minimal because there was no time left. At Belfast, the chief test pilot got in, Jamie got out and the three of us went flying.

The Shorts 330 flight deck has two doors from the passenger cabin, either side of the cabin – not the usual one in the middle from the aisle – and as you enter the flight deck, there is just enough space to stand by your seat. Strangely, the pilots' seats don't move forward or back, the rudder pedals adjust by winding them towards you or away from you, to achieve full rudder movement. I went first and it was all a bit of a blur and very average. Then it was Cami's turn and he was about as good as me. Cami is taller than me and a lot taller than Alan, the other co-pilot, but that didn't seem to be an issue at first. Then Alan got into the right-hand seat and completely omitted to adjust the rudder pedals to his shortened leg reach. No one noticed that he couldn't reach the rudder pedals to give him full movement, because Camille had wound them forward.

Before take-off, Camille was standing at the captain's door in the cabin and I was standing by the co-pilot's door, eagerly awaiting Alan's first take-off. We started accelerating down the runway and, just as we became airborne, the examiner throttled back one of the engines and the aircraft started to roll. Whether Alan, at the controls, managed to get off his seat enough to have a stab at the rudder and ended up pushing the wrong pedal, I have no idea. What I did know was that the aircraft was now on its side, skimming along the grass and we were aiming at a gap between two huge petrol storage tanks.

On the radio, in the middle of this crisis, all I could hear was: 'Shorts 1 [our call sign], Shorts 1, Shorts 1, are you in trouble? Shorts 1.' I thought: 'Trouble? We are dead!' Cami and I looked at each other and flung ourselves together, hugged tightly and shouted in a very loud voice: 'Goodbye!' Eyes tightly shut, we waited for the inevitable fiery end. But then nothing, and we slowly realised: 'We're alive!' At the final critical point, the chief test pilot had taken control and saved the day. He told Jamie: 'I'll sign them off, but you need to do some more bloody training!'

Third time lucky

When we got back to Glasgow, it was chaotic, with Operations and Engineering desperately trying to be ready for the first flight on the Monday. I wasn't rostered to operate the inaugural service and one of the other co-pilots flew that day. It was quietly, and embarrassedly, said to me that both pilots had forgotten to retract the undercarriage on one of the sectors. Not at all dangerous, but stupid. It was Loganair's first retractable aeroplane and it hadn't taken long to make this classic mistake.

Loganair had a new contract to operate the SD-330 on a service called Translink, paid for by the Irish and the Scottish Development Board as a support to tourism. It was going to fly between Belfast, Prestwick, Edinburgh, and Aberdeen or Inverness, and they were hoping that Americans flying into Prestwick would want to go to these cities. Loganair also had an oil support operation based in Aberdeen, where the Twin Otters flew up to Shetland with oil workers. From the very beginning it was a complete farce - there were hardly any passengers. All the pilots were Glasgow-based, as was the aircraft, but Glasgow wasn't on the route structure. The first crew would position the aircraft, empty, across to Edinburgh or Prestwick to start the service. Then, at lunchtime, the next crew had to drive themselves over in a company car to Prestwick or Edinburgh. Often, we'd be flying completely empty and it went on like that for the whole season.

Before I could fly the line and wear my uniform, I had to re-sit my instrument rating. Ever the optimist, I went down to Stansted – and failed again. This time I had forgotten that when doing a single-engine approach and go-around, you can only start the go-around procedure at 50 feet above minima and not one foot below. I went round at 60 feet and failed. One second later, and I would have passed. Undeterred, I did a little practice and went back and did the test yet again.

But it was not as simple as that - nothing ever is. Ruth and I went down to the CAA building at Stansted on June 8th 1979, and I went out with another candidate, who was an RAF Phantom pilot. We were at CAFU and the CAA examiner, Captain Shilkock, was taking both of us out on the same flight because we both needed to re-sit part of the test. I went first. The aeroplane was a Grumman Cougar GA7, registration G-REAT. It was an unusual-shaped aircraft and was the first Grumman Twin Cougar in Europe, a four-seater twin we had had to hire from the London School of Flying at Elstree airport. The chief flying instructor, Captain Mike McDonnell, was cross with me. He said: 'Scott, you're a good pilot. You've just got to get your head together when you do these tests. Whatever you're doing is not right because you're coming back with a fail when you shouldn't. I know, I can tell, you should pass.' This was to be my third test for the instrument rating and if I didn't pass this time, I would have to start the IR again. I was feeling a lot of pressure.

The re-test consisted of me flying on instruments with no visual clues. There were hardboard screens in front of me and to the side, so I couldn't see out. The screens were slotted in so the examiner, in the right-hand seat, could see out, and the idea was that I would fly down the ILS (instrument landing system) on one engine and fly a go-around at minima, simulating a no visual contact approach and go-around. The margin of error was fifty feet high, and minus nothing, so that was my tolerance. In this final chance with me in the left-hand seat, before take-off I was doing the mag check on one of the engines. One of the two magnetos was causing a large mag drop of rpm on one of the engines. I was messing around, trying to burn off any oil in the plugs by leaning out the fuel air mixture while under high power. We were sitting by the runway and the examiner was just watching me try to get this aeroplane serviceable to go flying.

This was a single-crew operation so he didn't help in any way. He was just watching me and telling me what we were going to do after taking off. I wasn't fully listening. I eventually got the mag drop within limits and I was thinking: 'Is it okay to get airborne or should I go back, cancel the test, and get engineers to look at the engine?' I didn't. We got airborne, and I got a bit confused during the departure from not really understanding the brief before take-off, and from the pressure of the whole pre-take-off. I put the gear (wheels) down, which was a mistake because they were not required at that point. Suddenly I thought: 'I don't need them down for what I am about to do.' So I brought them back up again. I shot the approach, did the go-around, and then the examiner said: 'I have control.' I had to climb in the back and another chap climbed in the front, we did whatever he had to do again, and then we went back and landed at Stansted for the result and debrief. I went into the office and the examiner said: 'Why did you put the wheels down?' I confessed: 'I just got a bit confused, sir.' Out of the blue, he announced: 'Well, you dug yourself out of it, you passed!' I said: 'Thank you, do you want a kiss?' He said: 'No, thank you!' That was me finished at last on the third attempt. The whole CPL/IR training and exams had been a traumatic, difficult and exciting time, but I had done it. I was a professional pilot – me, the boy the headmaster had said couldn't do it. I rushed out to tell Ruth the wonderful news.

Back to Scotland we went. There was relief all round that it was all done and dusted, but we still had a huge £30,000 debt to worry about, and I was only earning £3,000 a year. To make matters worse, Ruth couldn't teach in Scotland because she didn't have a degree at that time. I had surpassed all my own expectations. To me, my achievements were impressive. Within my career, I had done my instructor's rating and I had done an upgrading, although I was still likely to go and hooliganise round the UK if the urge took me, but if I did that, I now knew that Ruth would have something to say about it. I was starting to become aware that my antics in the air were having a direct effect on another person.

Up to that point, I was demonstrating to myself that there was no one else quite like me when it came to flying an aeroplane. When Ruth and I got together, lived in her flat and then married, she supported me wholeheartedly. Ruth has never ever been money-driven. She had chipped in her car fund money to help finance my flying after I had sold everything else of value I owned. Everything, that is, except my prized Breitling watch, which I had bought for £100 when a new one at the time cost £250. Don McGillivary, from the Glasgow Flying Club, had his own helicopter called a Brantly, a tiny two-seater with a petrol engine that sounded like a demented wasp. He took me up in it a couple of times and I had a go flying it. It was incredibly sensitive on the flying controls. He owned an impressive Breitling Navitimer pilot's watch, and he said: 'I can't use the damn thing flying the helicopter because I can't let go of the controls to look at it.' There was no autopilot and a helicopter is not stable like an aeroplane.

Sadly, after I had bought the watch, Don was killed in a helicopter accident. He had achieved his CPL/H (commercial pilot's licence helicopters) and bought his own Bell Jet Ranger, a jet-engined, six-seater commercial helicopter. He got a contract with electricity companies to fly along the power cables to check for damage. He would fly, just above the pylons, with an observer to check the cables. He must have loved doing it because he didn't need the money.

One day, he had an engine failure. The Jet Ranger had a single turbine engine and he was flying quite low at the time. If you have an engine failure in a helicopter, it is best to be either very low or fairly high, so you can auto rotate by diving down and letting the energy build up in the blade. At the last minute, it is possible to convert that energy to lift. If you are between the very low height and the auto rotate height, you are out of luck. That's why a helicopter is always very vulnerable in a low-level hover unless it has two engines. That's also why you don't often see helicopters going straight up: first, it takes a lot of power, and, secondly, at about thirty feet you get into a very vulnerable area where if the engine stops, you are in big trouble. Helicopters tend to go forwards after lift-off because they are building up speed they can convert into lift/energy. They were very low and slow, and they crashed to the side of the wires. The observer and Don both got out of the wreck, but Don's back had been broken. He walked a few feet and dropped and died. The observer survived.

Ruth was quite happy to move with my career. As far as we were concerned, we wanted to be together, whatever that meant. Of course, it meant paying the debt off and living wherever we had to for my work. In all the years, she has never been any different. When I had an offer from Cathay Pacific at one point, I said to her: 'Do you fancy going to Hong Kong?' She said: 'Not really.' But she would have gone had I needed to. Even when I joined Loganair and we moved to Scotland, she knew she couldn't teach. She could have said: 'I'd rather go to Norwich and you can fly for Air Anglia.' But she didn't. She put herself and her career completely second to mine without saying as much. All along, she has been a fantastic stabilising influence on me.

Loganair was only a little airline but it was still quite a formal operation, and I was learning all the time. I had done the manufacturer's course and went back and did some safety training with Loganair. There was still structure to the teaching and learning process. Nobody had ever shown me a fire extinguisher before, and certainly not how to tell its condition and how it worked. The safety training was done simply, but properly, unlike any flying club I had been to. There was a handling exercise, once a year, in the aeroplane because there were no simulators for any of their planes. After the handling part, which might include stalling and steep turns, we would join an airway and fly down to Prestwick, and fly various instrument approaches, just as I had done with the CAA on my initial test. Another yearly test was the line check, which involved a training captain sitting with you and observing during a normal line flight with passengers.

Our move to Scotland seemed to have happened quite quickly. The chief test pilot of Shorts in Belfast, Jack Cumming, had passed me to fly the new aeroplane. This was July 1979, and Ruth and I had been married for six months. I had passed all my CAA exams and for a while we lived in Cherry Bank Road, my auntie's council flat. Then we rented the house of my pal Robbie Hunter, who flew Aztecs for Air Charter Scotland, in Bridge of Weir, to the west of Glasgow. Robbie moved to Libya to fly dictator Colonel Gaddafi in his private French Dassault Falcon 20. Of course, I was concentrating on paying back as much money as I could to the bank. I was still paying back my RBS loan years later. I always felt that, of all the people in life one should avoid hacking off, one was your bank manager, and the other was your mortgage company.

Workwise, we were doing seven sector days, day after day, between Prestwick, Belfast, Edinburgh and Aberdeen. We had company cars so we'd pick up the plane at either Prestwick or Edinburgh. We were whizzing around seven sectors, and then back home. At night, the aircraft ended up in Glasgow. The Shorts was a pretty uncomfortable aeroplane and was known as 'The Vomit Comet', first because it wasn't pressurised and couldn't climb above the bad weather, and secondly, because it was a big slab of an aeroplane and didn't ride the turbulence very well.

Shorts had made the Skyvan before, which was a similar shape and smaller, and Loganair used to have one of these. I unloaded it many times when I was a baggage handler. The SD-330 was bigger, carrying thirty people, and needed two pilots and a single cabin crew. It was spacious inside but a bit uncomfortable in turbulence. It had an incredibly high wing, flat sides, and was an ugly-looking thing with a twin boom at the back. The 360, which came out later, had a single tail and I was told it rode the turbulence better, although I never flew that. Because the 330 was so new, the first off the production line, we were requested to take it to air shows. At the Prestwick Air Show, we flew it in and did a very low run down the runway and a steep pull up at the end, then parked it so people could look at it close up. It was a great opportunity for Loganair to show off its new aeroplane and the new service, Translink. Ruth flew with us and we had a great day. The highlights were a low-level pass by a Vulcan bomber and the Harrier jump jet bowing to the crowd. The trouble was there weren't many passengers who wanted to fly between Belfast and the Scottish airports. The whole thing was just a waste of time and fuel. It was a shame but, like most things in aviation, it failed.

I was still a co-pilot, a second officer, and I couldn't get a command in the Shorts because I didn't have an APTL. Out of the blue, I was offered the opportunity for a command and I suddenly went from one stripe to four. I was getting a command on a Britten-Norman Trislander, a three-engined aeroplane with 18 seats. I had never operated around West of Scotland as a co-pilot in Loganair, whereas the other guys had. The other people promoted had been co-pilots on the Trislander, or the Twin Otter, and had flown around Barra and Skye and the other Highlands and Islands destinations. I had been on the Shorts, bouncing around Aberdeen and Prestwick, flying on airways and on instruments, much more like normal airline flying. We would occasionally fly visually. At that time, I was flying with Ian Wilde, who joined Britannia before me. I later lived with him for a short while, when I first joined Britannia, and he still lives in my home town of Olney.

I was in the crew room at Glasgow, on one of my first days in Loganair, wearing my uniform with my one stripe on my arm. This chap who looked about a hundred came in – although he was no doubt only about sixty. This was Jamie Alderman. I said: 'Oh, hello. I'm new, sir.' He said: 'So am I, so am I.' He was in his captain's uniform and under his wings he had two rows of medal ribbons, which included the DFC and Bar and AFC and Bar. Jamie was a Trislander captain, and that's all I knew about him the first time we met. Soon after that, people would tell me: 'Oh, Jamie started Britannia Airways.'

It turned out that, far from being 'new', Jamie, at the age of just twenty-one and flying a Lancaster bomber, had led the last 1,000-bomber raid on Germany, towards the end of the Second World War. So here was a decorated war hero, looking years older than he actually was, and with ash down his uniform, humbly standing in the crew room with a group of captains. And he was first to say: 'Hello, and nice to meet you.' Jamie was a lovely chap – as long as he liked you. You had to be a wee bit punchy, otherwise he'd eat you alive. He is one of the small list of men who have stood out and influenced me throughout my life, both professionally and privately. All of them taught me necessary lessons or showed me guidance, and without them I would have achieved, and striven for, less. I would always call the captains 'sir'. I wasn't too sure of the protocol to begin with, and it took a while to get the hang of it. Eventually, one of the captains, a bit grumpy ex-Navy pilot – stuck in an aircraft carrier for years, you would get a bit weird – said: 'You don't have to call me sir.' I was always respectful because I believed they deserved my respect.

Local knowledge was important on the west of Scotland routes and I didn't really have much, as I had been flying the Translink service. The highlands and islands were very different: from a flying point of view, it is a different environment from flying down airways. Most places didn't have any air traffic or let-down aids. There was an NDB (non-directional beacon) on Skye, which was owned by Loganair, and only switched on when you were due to arrive, but that beacon only let the needle on my ADF gauge point to it. So I had to work out the bearing and distance, and these were worked out by flying over the beacon and starting my stopwatch, then flying outbound for a set time at a set speed. I had to include a time calculation for wind. You would let down to a safe altitude, then, at the calculated time, turn inbound and descend to minimum descent altitude. Not easy and not accurate, so the procedure couldn't get you very low, about 600 feet. It was another reason to stay visual and under the weather, rather than fly up, on airways, and try to get in using the beacons.

Whenever possible, we would fly up the west of Scotland low level because often the weather in Scotland was rainy and miserable. Visibility at very low level was often good because the weather lifted as it hit the Scottish mainland. The cloud base might be 200 feet but you could fly underneath it, and stay under it with good forward visibility. It was a different sort of flying, great fun, and a bit like the Dambuster raid but with no one shooting at you. We would fly up the Crinan Canal to get over the long strip of land running down to Campbeltown. Then, out over the Corryvreckan Whirlpool to the north of Jura. I had done some low flying around Mull and Loch Lomond when I was a private pilot, but that was just for fun and not trying to get somewhere under the weather.

I started the Trislander ground school, which was mostly home study with the odd chat with a training pilot or an engineer. Basically, they said: 'Learn it – you're booked in next week for your test.' It isn't a very complex aircraft and I had already completed all the general subjects that covered the type. The Trislander was a three-engined version of the twin-engined Britten-Norman Islander, built on the Isle of Wight. Both types have a high wing, like a Cessna, and a fixed undercarriage. It had nine rows of bench seats, with doors like a railway carriage along its length. It was powered by three 260HP Lycoming engines. You couldn't quite walk under the wing, unless you stooped a little. Two engines were mounted on the wings with the undercarriage's long oleos underneath and small double wheels at the bottom. The third engine was mounted high up, on the tail. The Islander was designed for short take-off and landing operations into rough strips. They were rugged 10-seater aeroplanes and were used by Loganair up in the Orkneys and Shetlands, for inter-island flights. They were also used for the Loganair air ambulance service because they could get into a farmer's fields on most of the Scottish islands. Even at night, if there was an emergency, the locals would light up the temporary strip with car lights and, in some cases, small portable lights to get the aircraft in. The air ambulance pilots were always the most experienced captains in the company. Loganair, and BEA before, lost air ambulance aircraft with all on board trying to get to a patient in poor weather. The rules went out of the window, but it was always the captain's judgment whether to carry on or throw the approach away. That was always difficult if there was a pregnant woman in a life-threatening condition on the ground waiting for the air ambulance.

Loganair still has the shortest schedule flight service in the world. It is between Westray and Papa Westray in the Orkney Islands and it is a flight of about 60 seconds. The record is 47 seconds. The Trislander carried an extra eight people but it was not as good at short field performance. The biggest problem with the Trislander, because of its increased length, was that it could tip up on its backside. After you landed, the agent would have to come out and put a pole on the back to prevent it tipping up when the passengers and bags were taken off. The western island service was like a minibus service for the highlands and islands, and was subsidised like the ferry service to cover costs.

My first flight in the Trislander was over Loch Lomond to do a flying test, which involved initiating stalling and engine failures, and other upper air work exercises. Although I had all my normal engine instruments, if the training captain brought one of the engines back to idle, the rpm didn't reduce much due to the constant speed propeller, but what I would notice was the yaw and roll because of the asymmetric power. When the trainer brought the power back to idle on the middle engine, there was no yaw or roll. He said: 'Have you noticed anything?' I said: 'No.' After a good look at the manifold pressure of the middle engine, which was low, I realised what was going on. Another very useful lesson learnt. There was a mirror beside me, mounted high up just outside my window, which looked just like something off a 1940s

MG sports car, and with this I could see my third engine high up on the tail. Crucially, it would not tell me if the engine was failing as the propeller would almost always continue to turn. The only way to tell if the tail engine was running was to study the engine gauges closely. I hadn't been doing either. We climbed up into the airways and I completed my instrument rating by flying a few approaches into Prestwick on two engines.

Next, I had to do line training, which involved a week of flying the line with passengers, and another captain supervising. I was being cleared for single pilot operation. I would do everything and they would just watch me. On the first day of my line training, we went off to Machrihanish aerodrome by Campbeltown, at the bottom of the Mull of Kintyre. This was an RAF airfield with a small passenger terminal away from the military one. I had not operated into any of the airfields the Trislander flew into, except for Glenforsa on the Isle of Mull. I would be operating into Islay, Tiree, Skye and Barra beach. We also went up to Stornoway and the Orkneys.

At the beginning of the week of line training, carrying passengers, something happened which was challenging, scary and potentially life-threatening. On my first sector, I was flying to Machrihanish, the RAF base. In bad weather, ATC (air traffic control) would give an SRA – surveillance radar approach – on to a PAR (precision radar approach). The PAR was something the Civil Aviation had stopped doing many years before, but the RAF was still doing it. On an SRA, ATC would give you headings or a turn of a couple of degrees, left or right, to keep you on an inbound track. On a PAR, they could also monitor you in the vertical plane and advise you to increase or decrease your rate of descent. I knew I was under radar control and being radar vectored to the west of Arran with a view to me coming down between Arran and the Mull of Kintyre. I would be turning, up Campbeltown Loch, to land at the far end.

My call sign was Loganair 411. We were flying in cloud and I could see that the captain who was with me was beginning to tune in a VOR. There was a VOR on the airfield but there wasn't a VOR approach available. At that moment, I was beginning to think he thought that something might not be right. It was never said, but the next minute, we popped out of cloud and we were in the hillside! We were in the bracken. I pulled back hard on the elevator and slammed the throttles fully forward. We didn't have GPS in those days so I was relying on the radar controller. At the same moment, I could hear air traffic on the radio as I was trying not to hit the hill: 'Loganair 411, pull up, pull up!' I pulled up hard, almost into a chandelle (where the pilot combines a 180-degree turn with a climb). The aeroplane, with almost no flying speed, just fell to the right into the valley and the loch, but not before the landing wheels had almost brushed through the bracken at the top of the hill, inches from disaster. As I was visual with the loch and the runway in the distance, I continued and landed on the westerly runway, in stunned silence.

Taking command

After a short time, I taxied to the terminal. The passengers didn't know how close they had come to their deaths. Loganair Trislander passengers were used to being bounced about, it wasn't a subtle operation. They would often have animals on their laps – sitting with sheep! – so whatever you did to them, they weren't too fussed. The aircraft was undamaged although we had just about hit the hill. I don't know what air traffic had been saying as we taxied in and the other captain didn't say much. We picked up the new passengers and went back to Glasgow. When I got back, the fleet captain passed me in one of the buildings and said: 'The RAF have just been on the phone and said that you declared yourself visual.' We had been in cloud, so this statement was devastating.

I went home that evening and went through everything with Ruth, Mum and Dad. What had happened? 'Did I declare myself visual?' The conclusion I quickly came to was that if I did, I shouldn't be flying. I went back to work the following day for the next phase of my line training, this time with another captain. He flew with me for the next couple of days, and I had more time to reflect on the Machrihanish incident. When I returned from the next day's flying, I quickly realised that this thing had not been resolved.

The fleet captain said again: 'Oh yes, the military have confirmed you declared yourself visual, and they are sending us the tape with the RT conversation on it.' More than one person who had spoken to me about it had said accusingly: 'You declared yourself visual – we've got the tapes.' On Friday, after the last flight, chief pilot Ken Foster, an ex-military chap and a bit of a bumptious sort, called me into his office. He was seated behind his big desk and I was standing, awaiting my fate. At one point, Foster had been so full of his own self-importance that he had wanted pilots to salute him at every exchange but it never happened. He was my boss and he held my career, and future, in his hands. I hadn't exactly been marched in but I'm pretty sure he would have arranged that if he could.

'You were,' he started, rubbing more salt into my wounds, 'going to be our first non-military captain.' The word 'were' hung in the air like a bad smell. This wasn't good. Actually, a chap called Davy Dyer had already been promoted to captain and he had come from a similar non-military background to me. At this point, I interrupted him because either I was not going to be promoted or I was going to be fired, so I had nothing to lose and I didn't want things to get that far without me saying a word. I asked: 'Is that the tape?' He said yes. 'Well, can you play it?' With a grunt, he reluctantly played the tape and what you could hear coming out of the machine was: 'Loganair 411, visual, right base.' This meant that you were calling visual, you were finishing with radar, and you could see the runway. 'Right base' meant there was the runway off to your right, and you were at base leg, before you turned right, on to the final approach to land. The tape player said: 'Loganair 411, visual, right base.'

I was alarmed at what I was hearing, so I said: 'Play that again.' He did, and I said, with every ounce of confidence I could muster: 'That's not me, that's Davy Dyer.' There was no more to be said. The only way to deal with this was proof. I walked straight out of his office, found Davy, brought him into Foster's office and told him to play the tape again. Again, we heard: 'Loganair 411, visual, right base.' I looked at Davy and asked: 'Davy, is that you?' He said: 'Yes, that's me.'

It started to become clear, even to the accusing Captain Foster. Davy's schedule the previous day had been the same flight to Machrihanish as I had operated, and in the afternoon he flew to Islay. We were all doing multiple sectors and 'Loganair 411' was the call sign for the morning flight into Machrihanish. 'Loganair 412' was the call sign for the morning flight into Islay. Davy was making his approach into Islay at the very same time I was making my approach into Machrihanish. He was talking to Islay on a different frequency, so I couldn't hear him, but the RAF ATC, for some reason, was monitoring both frequencies. Davy, on his approach into Islay, called using 411 instead of 412, mistakenly using the call sign he had used the day before on his approach in Machrihanish.

So he called 'Visual, right base' to Islay, and the RAF heard it. Islay acknowledged to Davy but on another frequency, so I didn't hear anything. The RAF heard the transmission. If they thought it was me, they didn't acknowledge as they should have done, and left me to fly into the hillside.

I got promoted after all and I received a letter of apology from the RAF, just saying: 'We've shot the controller!' They were quite contrite. I don't remember filing a flight safety report. Today, the Air Investigation Branch at Farnborough would have been involved and there would be a formal investigation, it was that close a call. If I had hit that hill, I still don't think they would ever have worked out the real cause. There was no flight data recorder, or voice recorder, on a small aircraft like the Trislander. They would have listened to the tapes and put the whole thing down to me. There would just be a plaque up there among the bracken, saying: 'Trislander, crashed here on the 7th January 1980, RIP.'

I came out of that fracas a bit worried about how to avoid this sort of thing happening again. I'd been told once about BALPA, the British Airline Pilots Association pilots' union, which no one had ever mentioned formally at Loganair. No one I knew, inside the airline or outside of it, was in BALPA. I could have done with some serious help here. Someone to ask what I thought might have happened.

The story of my near-miss went round Loganair very quickly and soon after that, a lot of us had signed up to BALPA. Davy was the only other Scottish pilot at Loganair and he had come from my background. Frankly, I don't think anyone could have distinguished our two Scottish accents over the radio - they were not brilliant radios - but I could tell, and Davy could tell, the difference. Pilots still use the wrong call sign, or respond to the wrong call sign, because every day they are using different ones.

The monitoring of the two frequencies was stopped immediately by ATC. The worst mistake had been ATC not acknowledging the call of 'visual' because at that point I would have said: 'What are you talking about? I am not visual.' In the same situation now, I would be more aware of my position – I wouldn't be leaving it to air traffic. There was, obviously, quite a long time with nothing being transmitted and I would ask for a radio check, to make sure the radio hadn't failed. My instruments couldn't tell me a lot; I was on their radar. The captain with me was fiddling with the VOR, possibly trying to see what radial we were on, but that would not have been enough on its own. One pilot couldn't be doing that and flying at the same time. There was no autopilot and you couldn't just let go. So I was in the hands of the controller. That episode exposed the weakness of a single crew and of that sort of approach with no back-up. If I was doing it again, I would be thinking: 'I'm going below safety altitude, I reckon I've got three or four miles to go so, in a minute, I'm expecting a turn.' After a minute and a half, I'd be on the radio saying: 'Just confirm…' If at that moment I thought it had gone awfully quiet, I'd be putting power on to get back up to a safe altitude, and then be back on the radio, asking: 'What happened?'

Nowadays, if you got that low on an approach, your terrain warning would start shouting: 'Whoop, whoop. Pull up, pull up!' But a Trislander is no more sophisticated than a Cherokee or a Cessna 150. It had basic flight instruments, an ADF and an ILS, but that was it. It still had carburettors on the engines, it wasn't even fuel-injected. The carburettors were prone to ice building up and eventually the engine would stop. You could blow hot air through the carbs to melt any ice but it was yet another lack of sophistication that kept the workload high. The leading edge of the wing had boots - like a rubber mat through which air would be blown. That would inflate and crack any ice which formed on the leading edge of the wing. That was the theory, anyway, but these aeroplanes weren't really designed for Scotland in winter.

One of my first flights as captain was to the Isle of Skye, and Skye had quite a demanding instrument approach. It had a beacon, which Loganair owned and was switched on only for its aircraft. No one had told me that I needed a check-out there, or a clearance flight by a training captain. I was just told: 'You're going to Skye.' I got in my Trislander and off I went. I was climbing out of Glasgow and the weather was foul, so I had to climb into it. Eventually, I was at about 3,000 feet above all the terrain. I couldn't get much higher with the old Trislander, and I was starting to ice up. The windscreen wasn't heated but there was a small screen, about the size of a shoe box, which just leant against the Perspex windscreen directly in front of the pilot, and that was heated. My pitot tube was heated, so that gave me the source of my air speed, and I also had the de-icing boots on the wing leading edge working to break off the ice. All I'd really managed to do was push the ice off the leading edge of the wing.

Ice was building up on top of the ice pushed out by the boots, and my boots were moving in and out between the ice and the leading edge. As a result, my wings started to ice up. Then my pitot tube heater failed, and I had no air speed indicator. I could see the pitot tube outside and it looked as if someone had stuck a big snowball over it. The aircraft was slowing down, as it was getting heavier with the ice, but, more critically, the ice was breaking up the airflow over the wing. The aerodynamics of the wing had changed and I was losing airspeed. By now, I was about fifty miles north, north west of Glasgow, over the hills. I thought: 'I can't do this. There's no way I can do an approach. I've no air speed and I need a big long runway, and help.'

I turned back for Glasgow, slowly descending into warmer air as I was clearing the Highlands, and the safety altitude was getting lower. The pitot tube was still frozen. I called Glasgow and said: 'I've got no air speed indicator and I am icing up and I need radar vectoring on to the ILS.' I flew down the ILS but because I couldn't be sure of my airspeed, I couldn't lower any flap, so had to come in flapless and at a higher speed. My heated panel on the windscreen had also failed and I couldn't see out forward, so I opened the little DV (direct vision) window, by my side, to see the side of the runway. I landed okay, but I had to tell ATC that I would have to stop on the runway, shut down, get out and scrape the windscreen clear before I could taxi.

I was checked into Skye a couple of days later. In those days, this was the fantastic learning environment for young pilots which we just don't get now. We flew underpowered aircraft, with no pressurisation, no weather radar, and no GPWS (ground proximity warning system). Today, people will come out of training with a commercial licence but they've never been upside-down, they've never been frightened, and they've never had to dig themselves out of a big hole, which they had dropped themselves into.

We used to do a daily flight to Islay. It was an old RAF airfield used by aircraft ferrying across the Atlantic from the US during the war. Islay had a proper ATC service and a control tower. There was one non-precision approach there, and again you had to use the beacon and your stopwatch to fly the approach in poor weather. It was a demanding approach as it was single pilot with no autopilot. The good news was that in the west of Scotland the weather seldom had very low cloud, poor visibility and high winds all together – it was usually just two out of the three. BEA had lost an ambulance flight there when one of its de Havilland Heron aeroplanes hit the hills during an approach. Some time after I left Loganair, a Twin Otter hit the hills there too. Quite a few people survived that crash, but the propeller decapitated the captain. In terms of modern aviation, I was still in a different world. Someone told me recently that Loganair had lost about five aeroplanes over the years. This is due to flying in a demanding environment, in very unsophisticated aeroplanes with little or no letdown aids – and often with low-experienced pilots. This was long before GPS and the modern aircraft they fly today. The upside to all of this was: boy, were you learning quickly. In that year, 1980, Ruth and I bought our first house and we were in our new home at the time of the birth of our first daughter, Inga. We had to buy things for the new baby. Mum bought us a pram, we had a cot, but we had no carpets in the house. We had some curtains and we had a phone. Loganair paid for the phone, because they used to phone me in the morning on a wake-up call. Inga was born on June 21st 1980 and Ruth's Mum and Dad had come up for the birth. I remember being so excited and smoking a cigar with some of the chaps in the crew room at Loganair to celebrate. We called the baby Inga because I'd been on detachment to Orkney earlier in the pregnancy, and I had heard the name there.

I'd never been to Orkney before, and it was an amazing place. There are still the remains of partly sunken ships beside the Churchill Barriers. A German submarine, U-47, penetrated the block ships and sunk the Royal Oak. You can still see the oil coming from the Royal Oak, which went down there with more than 800 hands. There was a constant oil slick over the wreck. I would pass over the slick coming in to land. I was sent there for a month when another captain was taking leave, replacing him and his Islander with my Trislander, which couldn't get into all the airfields. I was met in Wick by the captain I was replacing from Kirkwall, and he was going to give me a quick briefing, including a few tricks of the trade for that environment. His briefing was, basically, if the weather was good, I should fly visual. If the weather was bad, I should fly instruments – and if it was really bad, I should stay below it all and stay visual. As we were going along quite low, he said: 'Right, here's the island now. You don't take the first little river, you take the second one. And then you pull up and there's the runway, and then you land. Bye, see you in a month!'

I flew into Inverness one day, coming in from the north, and there was another Trislander coming in from the south with Ruth and Jill on board. As the wind wasn't very strong, I would have liked a straight in approach on to the westerly runway 24, so I called up, and said: 'Is there any chance of a straight in approach?' The reply was: 'Negative, the other runway's in use.' The other Loganair aeroplane called up, just after me, and said: 'Good morning, Inverness, this is Logan 235, request radar vectors to Runway 24.' The controller told him: 'It's not the runway in use. We are using 06.' The other pilot was immediately back on the radio, saying: 'I didn't ask you, I'm telling you, that's the runway I want.' After a short pause, ATC meekly replied: 'Oh, Roger… er, you're clear for that runway.' ATC then asked me if I still wanted runway 24, and I said: 'Yes.' I landed first, as I was ahead of the other Loganair aeroplane. After I parked and the other aircraft had arrived, it turned out that it was being flown by none other than old Jamie Alderman. He came up to me and said: 'Son, don't ask. They're there to look after us. Never forget that. Tell them what you want, not what they want.' That was the first time I came across the authority of the captain. I guess you learn exactly what command is all about, leading a 1,000-bomber raid on Germany.

Some time around then, I was doing a flight to Barra, where you land on the beach with a few poles to indicate the beginning of the three official runways. Due to the high tides, the flight was scheduled at a different time each day. Sometimes the beach would still be wet, especially if there was an easterly wind blowing the water up the beach, making it quite a demanding place to land. We would fly in and out with the Trislander but this aeroplane wasn't ideal for Barra. It had a fixed gear but little wheels, which tended to sink into the sand. Each wing had two small wheels and then a wheel at the front. A Twin Otter, on the other hand, had great big wheels and also had reverse thrust so you didn't have to use the wheel brakes. The way to do it, when the water was up high on the beach, was to do a curving approach and a curving landing.

There were no letdown aids at Barra and, before I left Glasgow, I would phone Kate Macpherson, Loganair's station manager, and she would say 'Och aye, you know, the beach is fine', 'it's a bit damp', 'you'll be all right' – or 'don't come'. We would fly over Tiree, where there's a VOR beacon, and we would fly outbound along the radial towards Barra. The VOR had a DME (distance measuring equipment) and that meant we would have a distance from Tiree and a bearing. Then we would fly out on a radial towards Barra, 37 nautical miles away. As we were descending, we would lose the beacon so we had to descend within 10 or 15 miles of Tiree and then run in to Barra, visually, at low level. That meant descending fairly smartish while I had a strong signal from the beacon, so I could get below the weather.

On one day I had a co-pilot – I was usually single crew – and I had been told at the briefing that morning that there was a naval exercise going on to the west of Scotland. The weather was pretty duff. Any time we could fly visual, we were told to do so because it saved airways charges. I would often get airborne out of Glasgow, cancel IFR and go visual because the company wouldn't incur en-route charges. I would do the whole route visually, as you would in a Cherokee. On this particular day, you couldn't.

We got to Tiree and were outbound on the radial towards Barra. We descended to get under the cloud and the minimum you could go down to was a few hundred feet. I knew we were over the sea and there was nothing there, so we got down to a couple of hundred feet, 12 miles out, still with a good signal and we were in cloud. I could see bits of sea below us through broken cloud so I said to my co- pilot, who was the pilot flying: 'Go down another 50 feet.' This young lad – I was young, but he was even younger – dropped down and I could then see the sea quite clearly. I said: 'Down a bit more, down 100 feet.' It was possible to see that we were coming out of the bottom of the cloud and visibility was improving. We got to about 80 feet and I could see forward clearly, just clear of the cloud. We were dropping lower and lower. Legally? No, but if you didn't operate like that, you never got into these places. At 15 miles out, we got down to 60 feet, 50 feet, suddenly we were out of the weather, and you could see for miles. And at that very instant, a huge submarine – obviously doing some fancy manoeuvre – came crashing out of the water, directly in front of us with the bow appearing out of nowhere at a huge angle. We wouldn't have hit it but my first impulse was to yell: 'Aaaaagh!'

The co-pilot pulled up sharply and went straight back into cloud. We had to climb up, go back to the beacon on Tiree and start the whole letdown again. The sub had come up from the right to left and was almost certainly doing a crash surface manoeuvre. I would love to have been on the sub when they popped up, only to find an unidentified aircraft on top of them. I bet we were the only civil airliner that has ever had to pull up for a submarine. Anyway, we landed safely and the passengers were none the wiser. Coming back was always easier because I could climb up, knowing I could do a precision approach into Glasgow. Jamie and I were both captains on this route, flying the Trislander. Jamie liked a drink, as a lot of pilots did in those days, but he wouldn't talk too much about the war. As I got to know him, he would tell me little things. He said he left the RAF after the war and was looking to get a job in the airlines. He was lucky enough to have flown the Lancaster and in those days everyone wanted pilots with heavy, four-engined experience. He dug roads for a while to earn money but eventually joined KLM and flew Constellations for them across the Atlantic. He was one of the original Atlantic Barons, a group of pilots who used to have their own flats in New York. Despite working for KLM, Jamie never moved to Amsterdam. He lived in Glasgow and would pick up the KLM service when it arrived at Prestwick, then fly it over to America. Legend had it that it was Jamie who would get airborne, then say to his crew – a radio operator, a navigator, a flight engineer and two co-pilots: 'I'm changing into my pyjamas now and

I'm going to have a sleep.' One day, the KLM Constellation, with a different captain, failed to arrive at Prestwick. It crashed into the hills with awful loss of life. The weather had deteriorated and their charts were inaccurate and didn't show the correct height of the terrain and the electricity pylons. The crew did a go-around, but hit the hill and the pylons to the north of the airport. In those days, they didn't have all the pilot aids that could save you now. There wasn't the instrument landing system, and there was no GPWS as the very last chance of survival. You always need to make full use of aids like that when you have them. Unlike a South American B747 flying into Spain, which had screwed up its approach. The GPWS started shouting: 'Whoop, whoop. Pull up, pull up!' The last thing heard on the voice recorder, just before the crash, was: 'Shut up, gringo!'

The Connie crashed north of the airfield into the hills. They were not particularly high, but they certainly brought the thing to a sudden stop. Jamie did that run for quite a while but there came a time when the Dutch wanted to get rid of the foreigners and replace them with their own people. So he left KLM, possibly because he was approached by the person who was starting what became known as Britannia Airways. In those days, it was called Euravia, and it started in 1961. Jamie was always loved by his passengers. He looked the part of the old airline captain who had travelled the world, with his medal ribbons on his uniform, his cap always on his head and cigarette in his mouth. He oozed confidence, unlike us boy pilots. Jamie always chatted and joked with his passengers. When in cloud and with ice suddenly crashing off the propellers in lumps and banging on the side of the aircraft, he would just call out: 'It's only the ice for the gin and tonic.'

When I flew the Shorts 330, there was one cabin crew on board out of a group of about six. The 330 was the only Loganair aircraft requiring cabin crew. It was at Stand 21 that Captain Ray Valerio lost his passenger. A passenger asked one of the check-in girls 'Is that my captain?', pointing out of the window to Ray. He was older than Cami and me, but he only looked sixteen. She replied yes - and the passenger refused to board the flight. On one occasion, when Camille flew with me as my co-pilot, we lost an engine coming out of Glasgow in the Trislander. Because I was the captain, I took control and flew back in on the remaining two engines. It was no great drama but I still got a bollocking because the engine hadn't stopped altogether, it had lost power and I had closed it down. Chief training captain Alan Whitfield asked me: 'Why did you close it down? Why didn't you leave it at idle? Could it still have given you some power?' I replied: 'Well, I didn't want to damage it.' He said: 'If an engine is still running, even on low power, it's better than nothing. Because this engine was shut down, I've got to put a report in to the Authority.' I thought: 'Okay, there are different ways of dealing with this. Why not keep the engine going, even at low power, to keep the generator running?' Another valuable lesson learnt .

My smart Loganair captain's uniform

The chief training captain also flew the line occasionally, and the air ambulance.

One time, when he was based up in Orkney, the CAA were visiting and doing an audit on the operation. The captain from the CAA had come up to spend a bit of time with Alan and was inspecting the operation there. During the inspection, there was a call-out in very bad weather for the air ambulance. A young mother had had a car crash on one of the remote, outer islands and it was very serious. It was foggy and of course there was almost no medical help and certainly no hospital. When they got a call-out like that, they always had to balance the risk to the lives of the doctor and the nurse on board with the need for urgent help. Risk or no, Alan was going and the CAA man said: 'Can I come along?' Alan said no, but the inspector insisted. 'I would just like to come and see, not an official observation, but just come with you.' Alan relented: 'Okay, but you say nothing.' I heard this story afterwards, from the CAA inspector. He told me they got airborne or, more accurately, they got up off the ground and levelled off and went at low level over a fence. Suddenly, as they passed the beach, Alan cranked it round then levelled the wings. The inspector thought: 'We're only at twenty feet and there's almost no visibility.' Alan was looking down out of the side window and suddenly cranked the aeroplane around again, levelled the wings and landed. When the local ambulance people were putting the young lady into the Islander, the CAA chap said to Alan: 'How did you do that? There were no navigation aids and no visibility to see where you were going.' Alan replied: 'It's the tide. I did the first turn off the beach, then turned where two sea currents meet.' Even flying in the fog, he could navigate on local knowledge of the sea conditions. He knew the point where there was always a line on the water visible from the air, and he knew he should execute this last turn exactly over that spot on the estuary. They got the crash victim on board, Alan climbed straight up into weather and flew airways back to Aberdeen and the hospital. You can never teach that kind of skill. It was real frontier stuff. The locals in those remote parts of Scotland thought very highly of the Loganair air ambulance pilots. Helicopters now operate the service but they were not capable back then. In a helicopter, you can stop and hover, or just land if you are not sure where you are. Try that in an aeroplane!

I always wondered why experienced pilots from military or civilian backgrounds joined Loganair. The usual answer from the Navy and RAF pilots was that they fancied flying around Scotland, often at low level, rather than flying for the airlines. But your salary was almost always influenced by how many passengers the aircraft you flew could carry. The theory was that the more passengers, the more profit it would generate, and of course Loganair's aeroplanes were tiny. By the time they were fed-up fighting the weather and had realised that there would be a huge difference in salary, it was too late. To move on, they would have to join the jet operators as the most junior co-pilot and hope for an early command, something they should have done straight from the military. The other group, the experienced civil pilots, for the most part didn't have, or couldn't get, their ATPLs. Either that, or they didn't have the confidence to fly jets. Alan Whitfield wasn't like that. He really enjoyed the environment and the whole Loganair operation and, by achieving the position of chief training captain, he had a good and interesting career.

I made the move because I was still young and because I thought it would be magnificent to fly a jet. The fact that it paid a lot more wasn't really the reason, but all this was to come later. At that stage, I was loving being a Trislander pilot. My perspective began changing as I spoke to other pilots from other airlines that flew into Scotland. Scotland was quite insular and, other than BEA, the only other airlines you would see might be an Icelandic B727, a BAC 1-11 coming from Ireland, and maybe a KLM DC-9. There would also be charter flights, such as the Britannia B737s, coming in, and Dan Air Comets and Caledonian BAC 1-11s from Gatwick. I wondered if I could fly a jet? My imagination was fuelled by Jamie's stories about Britannia Airways and his days flying the magnificent Lockheed Constellations across the Atlantic to New York for KLM.

Jamie had an amazing ability in the air. I flew with him a couple of times and on one final approach into Glasgow, he suddenly enquired: 'Who's flying it next?' The answer was: 'It's Camille and he's going to Islay.' He would be setting up the aircraft on the approach for Camille's departure. He would call up an approach frequency and he'd say: 'Hello Glasgow, this is Loganair 412. We're over Largs visual right base 23 Stand 21.' They'd say: 'Okay, Jamie!' He was hugely respected at Glasgow Airport by all the people who knew him. It was as time went by that I realised that, as a Britannia captain with the distinctive white cap, he must have been a very recognisable figure.

In those days, pilots going abroad were allowed to bring in just twenty-five cigarettes when they came back to the UK. Captains who brought thirty in got a right bollocking from Customs. Jamie would come in with 200 cigarettes plus a bottle of whisky, which you weren't allowed to bring in. He once arrived at Customs and they asked him: 'Anything to declare, Captain?' Jamie replied 'The usual' and the Customs chap said: 'What's that?' He said: '200 and a bottle of whisky.' The Customs guy was a temporary placement and said unwittingly: 'You can't bring that in.' That prompted a great hoo-ha, until someone took him quietly aside and said: 'It's okay, it's Jamie Alderman.' I guess when you have been shot at night after night with no real hope of surviving, you are not going to be fazed by some young upstart. Sometimes when he took off from Glasgow, he would get airborne in the jet and, at 1,000 feet, crank it over and fly directly over his own house. You couldn't believe someone could get away with that, you certainly couldn't today. I can hear the noise lobby complaining at the thought of it.

Fun and games aside, Jamie was still drinking. We would go to the Excelsior Hotel at the airport and Cami and I would join him for a wee drink in the evening. After quite a few drinks, we would make our excuses and leave him at the bar. We would both be off duty the next day. Jamie would appear bright and ready to go flying the next morning, an amazing recovery. The attitude to alcohol was very different in aviation then. I was beginning to think Jamie was coming to the end of his flying career. I don't know exactly how old he would have been, perhaps sixty-four, but I was with him when he did his last flight.

I was actually the commander and he was in the right-hand seat. We flew out to Tiree, and when we landed there, on the small terminal building there was a banner saying: 'Goodbye Jamie!' When the passengers got on board, he told them: 'It's my last flight.' He got in the right-hand seat and I said: 'You fly it back.' As soon as he got airborne – I exaggerate with the word 'airborne', we were only just off the ground – he turned the aeroplane and beat the airfield up, and I mean spectacularly beat it up. Not content with that, he went through a couple of farmyards as well. It was amazing. On the way back to Glasgow in the Trislander, which was not exactly swift, he had thrown it around as if it were a Cherokee. As we got back to Glasgow, he asked defiantly: 'Shall we do a run in and break?' This would involve coming in fast, as military aircraft sometimes do, coming down the runway fast, checking everything's okay and then pulling up to lose all the energy before coming back down to land. I, being the commander, had to advise: 'No, Jamie. I've still got a career ahead of me, you know. This is your last flight.'

He always used to land quite deep in the runway so he didn't have to taxi for too long. He would keep some power on the rear engine and come off the high-speed turn-off with the front wheel still in the air. We were told that we were parking in one of the remote stands, Stand 80, and there would be a coach for the passengers. We thought that a bit unusual. We only had 12 passengers and usually we would park at Stand 21 on the end of the pier. When we arrived, we found there was a marshaller to help us park, a coach for the passengers, various management from the airport, his wife and television cameras, and a big reception that had been laid on by his former colleagues in Britannia Airways. It was an amazing sight. The whole event was reported on TV and in the press the next day. Jamie went on to become director of the Red Cross for Scotland, a position he ran from a wonderful office in Paisley. We would sometimes go and see him for coffee and ask his secretary: 'Is Captain Alderman available?' If he heard us asking, he would come out and say: 'I am not a captain now. I have left all that behind.' But there was no way on God's Earth any of us would dream of asking for Mr Alderman.

Around the time Jamie finished flying, there was news of redundancies at Loganair. By that time, I had applied to join Britannia and I had made another application to join British Caledonian. I received a letter back from Britannia declaring that I didn't have enough experience and I needed jet time, so thank you very much. I recounted that to Jamie and, strangely, just a few months later, I received an unexpected invitation to attend an interview at Britannia. I went to Luton and was interviewed by Ted Barff, the chief pilot and another chap, both very senior men. I was sitting outside the office in the building which, in those days, was almost like a Nissen hut. I could hear another guy being interviewed and they were asking questions like 'What's the range of a VOR?' or 'How many miles out is a localiser acceptable to lock on to?' These were technical questions and I sat there thinking: 'Oh dear…' I walked in with as much confidence as I could muster and they said: 'So you're Scott from Loganair?' Their next question was: 'You know Jamie, don't you? How is he? … Anyway, thanks very much. Away you go.' I was offered the job.

Loganair was going ahead with the redundancies and I said to them 'I'm going in September to join Britannia Airways', giving them lots of notice. This was around May. I thought it was fair to tell them this, but of course I ended up shooting myself in the foot. When they announced the redundancies, I was one of them. I wouldn't have been on the list had I not said anything about leaving. I had been with Loganair two years and was being made redundant. I understood why they were letting me go. There was no point keeping me on for an extra few months and making someone else redundant. I didn't mind because there was so much to look forward to and my next adventure was already awaiting my arrival.

At least, that's what I thought. I received a letter shortly after from Britannia and a cheque for £888 fell out. Ruth announced happily: 'They've paid us already!' It wasn't a payment at all, it was a pay-off. They wrote: 'Don't come on Monday – you're redundant.' Even though they'd offered me a job and I'd signed a contract, they now didn't want me and the cheque was for a month's salary. I was twice redundant – at Loganair and now at Britannia. If Loganair were sympathetic to my plight, they certainly didn't show it. When I explained my predicament to them, they replied bluntly: 'Well, sorry, tough…'

Hero pilot, drunken friend

As far as Britannia were concerned, they had signed me on and they had signed me off. It wasn't long before I figured out that I had enjoyed a leg-up at Britannia because of Jamie. When I went to Britannia and mentioned Jamie's name, I could see quite a few people wince. The bigger characters, the captains, the strong men, would all say of him: 'Good lad. I like Jamie.' The quieter, more timid chaps, like some of the first officers, would go: 'Oh, he used to kick me in the bar.' Apparently, Jamie had a habit that if you weren't drinking enough or he didn't like you, he would just kick you. As a first officer, you couldn't really say to a captain like him: 'Stop kicking me!' It was like the old days of the surgeons: they thought they had the right to bully you if they chose. He used to say to me: 'You're a pilot, you get to the bloody bar and have a good drink and then you deal with the women. You don't piss about.' At that stage of my life, I would say meekly: 'Yes, Jamie.'

He and I were quite close, but what happened to him in Britannia at the end was not generally known by the newer guys. I was told by one of the old bosses that Jamie and another captain were in Tenerife for a few days. One of them was due to operate and the other was on standby. It was cheaper to keep aircraft abroad because parking and crew hotels were cheaper. Jamie and this other captain had been drinking solidly for days. Report time came and the crew were gathering in the hotel lobby – no Jamie. His first officer, who was unfortunately new and inexperienced, knocked on the door and, looking in through a fog of cigarette smoke, found Jamie and the other captain both the worse for wear.

I don't suppose the first officer knew Jamie well enough to know that at that moment he only had three options. He could throw himself downstairs and therefore ground the operation with a broken leg. He could break the aeroplane by finding, say, a great big important circuit breaker, pull it and snap it off. Or he could tell Jamie not to fly. That last option was never going to work because Jamie, for all his faults, had amazing recovery and half an hour later would be showered and downstairs. There wouldn't be a sign of the carnage just a short while before. It was unbelievable. Anyway, the first officer elected to say: 'We can't go.'

A similar thing happened when I was in Loganair, when I was a first officer and I had to dress a senior captain in his bedroom. We were night stopping in Inverness and the captain got very drunk. The next morning, the stewardess and I were ready, but no captain. I phoned his room but no answer. I went up and banged on the door until he opened it, looking dreadful. He had still been asleep. I dressed him and said: 'I'll put you in the aeroplane but you don't touch anything!' He didn't and I flew the Shorts back to Glasgow. The captain wasn't a very pleasant chap, but after this incident he didn't bother me again.

Jamie said to his first officer: 'No, I'll be all right, son, I'll be fine.' The first officer did the right thing and grounded the flight - and reported Jamie to the company, so the shit hit the fan. When the two captains arrived back a day or so later, as passengers on a Monarch flight, they both came off the plane the worse for wear again. He was hauled in front of Derek Davison, Britannia's managing director, or 'Father' as we used to call him. Derek had started Britannia with various captains, including Jamie.

Quite a few years after I joined the airline, Derek told me the story from his side of the desk. He described how Jamie had walked in and said in a confessional tone: 'Well, you're going to have to fire me this time.' Of course, that made it so much easier for Derek, as Jamie knew it would. Derek replied: 'Yes, you're going to have to go.' Derek also told me that Jamie had spoken to the first officer after the whole thing was done and dusted, and told him he had done the right thing in grounding the flight. The other captain didn't make it at all easy and tried to keep his job, but he was fired as well. I didn't know at that time why Jamie had left. I thought he had just retired at sixty, because until about 2006 that was the retirement age for pilots flying large aeroplanes. After that, pilots could go on to sixty-five. You could fly at Loganair until you were sixty-five because they were only operating little aeroplanes, and pilots were flying using their CPL, not ATPL.

I guess the CAA wouldn't have known anything about the incident. He went to Loganair and I met him for that first time, a year or two later, when he announced rather modestly: 'I'm new too!' That was not the first time our paths had crossed. A few years before I joined Loganair, I had been in Mull with my Jodel and had the nerve to borrow/steal with permission some fuel out of a visiting Trislander when I was short of fuel. Jamie was the Trislander captain. I have film footage of the Trislander coming in to land at Mull, and out came Jamie. I spoke about Jamie once to chief training captain Chris Clarke, one of the bosses of Britannia at the time, who was well respected and went on to found the CTC training school. CTC is one of the few training establishments where you can go in off the street, having never sat in an aeroplane, and come out eighteen months later with a frozen ATPL – all exams passed but without the flying hours – and with a bit of luck, join an airline as a first or second officer.

Chris told me that years before, when he was a co-pilot, he was going into Glasgow one night with Jamie and it was blowing a howling gale. Most captains would be concentrating on flying the aircraft, really working hard, and there would be no conversation. Chris said to Jamie he had been struggling with his cross-wind landings, and while Jamie was flying the aircraft in dreadful weather conditions, he started to tell him what he was doing and thinking: 'You want a bit of this, son – two degrees nose up, keep the power about 1.2 EPR, just gentle movements, don't over-correct.' Chris said he pattered it the whole way down, hardly looking out, looking over at him most of the time – and landed. Chris just sat there and thought: 'Bloody hell, how good was that?'

There was a small group of businessmen, led by Ted Langton, who arranged the funding for Euravia, which later became Britannia Airways. Derek Davison, who went on to be managing director and then chairman when I joined, was one of the original pilots and chief pilot. Derek was also a former bomber pilot. He started the airline with Jamie and Don Tanton in Luton with a couple of Lockheed Constellations. They were beautiful, four radial-engined long-range aircraft with a triple tail which they had bought from El Al. In the early years, they were doing any route they could get their hands on, including carrying cargo if necessary. Whenever they sent Jamie anywhere, he would always spend twice as much as any other captain. He would come back from somewhere like Hong Kong with a load of cargo. Where everyone else was told 'Fly out with a load and come back empty', he would do a deal and would make money for the company.

Part of the family

As I headed off towards new horizons, I reflected on the life I was going to be leaving behind. I had been very happy at Loganair. I would drive to the airport in the morning, go up for a meteorologist briefing and the Met man would say: 'Where are you going today, captain?' I'd say we were off to Londonderry. He would have a chart out and say: 'Well, you're going to get a fairly clear run if you're going to go low level. As you come round this part of the island, you know, coming off the bottom of Arran, there'll be a bit of haze but you'll break out of it.' Their knowledge, and their local knowledge, was unbelievable. We would have our own briefing by an actual human being.

Sometimes the Islay and the Campbeltown flights would leave within minutes of each other, so you might find that the other skipper arrived at the same time and there would be a couple of us at the briefing, but it was never a group briefing. It was always flight-specific. We'd be briefed by the flight information people and they would give me all the NOTAMS (notices to airmen): what's broken, what's not broken, what's serviceable, and what taxiways aren't working. Then I'd go into Operations, where they'd say: 'Morning, Scott.' My aeroplane might be over at Stand 21, or it might be in the hangar, and they'd give me an update on the serviceability of the aircraft. Then I'd go to the aeroplane, check its tech log, start it, and taxi over to the stand. All the engineering and walk-rounds would be done by engineers. I would have a quick look round myself and just say what fuel I wanted.

They had standard fuels for all the islands, so it was all done for me. We would put our passengers on, and away we went. It was a very sociable place. The first flight of the day was an eight o'clock departure and I would be finished by mid-afternoon. In January 1981, I should have been going straight to Britannia but it looked as if this was not going to happen. Before even starting there, I had been made redundant. I had a pension from Loganair but of course there wasn't much money in it. They asked me if I wanted some cash or, alternatively, I could have the chief pilot's company car, a Triumph Dolomite. I chose the car. But I still needed a job.

In Aberdeen, there was a company called Air Ecosse that was doing the oil work with Twin Otters, Trislanders and Bandeirantes. They were never seen as a great company to work for - Loganair was regarded as a much better outfit. I called them up anyway and said: 'Are you looking for any pilots?' They said they might need a Trislander pilot, so the next day I drove all the way to Aberdeen. When I got there, they said they could offer me a part-time contract as a captain on the Trislander. I never told them about Britannia. I told them: 'I'll get back to you. That is very kind.'

I drove back and the next day I went into Loganair and asked 'Is there anything happening?', to which they replied no. I said: 'Well, I've been offered a part-time job as a captain at Air Ecosse.' They suddenly changed their tune: 'Och, well, maybe we can use you as a part-time captain.' I went back to Air Ecosse and said: 'Loganair can offer me this.' They then said: 'We can offer you a full-time contract as a Trislander captain.' Loving all this, I went back to Loganair, told them of my latest offer, and they said: 'Oh, for goodness sake. You can have your job back.' The only downside was that it was to be as a co-pilot on the Shorts, but they said I would keep my captain's pay and rank. That was just fine by me. I went back to Loganair on my old salary and I could hang on to the car. Above all, I could pay all my bills.

I was back as a captain but flying as a co-pilot, and I still hoped I might be able to go to Britannia one day. To do that, I would have to re-apply but I knew that Loganair knew that, so there was no secret about it. Loganair put me on a Twin Otter course but I never flew it operationally. I did all the checks but they thought they might get rid of the Twin Otter, so they said: 'Go on to the Shorts.' It was lovely because Camille was now a captain on the Shorts. He had done his ATPL and I hadn't, so I flew with him as his co-pilot, just as he used to fly with me as my co-pilot. I began flying again with a lot of the old captains I had flown with before, and that went on from April 1981 until the end of that year, when I left for good.

I was still doing the Edinburgh-Belfast-Glasgow-Inverness runs, the same old stuff. By now we were also flying to the Isle of Man, so we were expanding, but it was still basically day flights. In the summer of

1981, they asked me to fly the Trislander as captain and the Shorts as co-pilot, which I was happy about. I was flying in the right-hand seat of the Shorts and the left-hand seat of the Trislanders, which was great. I did that right up to the time when my new course came through with Britannia. It was due to start on January 18th, 1982, and this time I thought: 'I won't say anything to Loganair until I have to.' My last flight with Loganair was on December 24th as a co-pilot in a Shorts to Inverness.

Having said my goodbyes to everyone at Loganair, I could finally look forward to my arrival at Britannia. I needed to know the lie of the land at Luton before I arrived as the fresh-faced new boy, so I decided to pay a visit to the site the day before my official start date. I met up with a kind security chap named Sam, who gave me a very useful recce around the whole place so I knew exactly where I was going. There was a mixture of ex-military, a couple of people with my background, and the rest sponsored by Britannia, who had completed the full-time eighteen-month course. Everyone was going to be trained to be a co-pilot and first officer on the B737-200.

On the ground school we were paired off in anticipation of the simulator training. The ex-military pilots were a mixture of heavy jet people and some fast jet people. I was paired up with Dave Balman, who was about my age but hadn't been flying as long. We were given our position on the seniority list, which is all-important in an airline. Seniority affects your promotion, redundancy, leave, base moves and change of aircraft type. It is core to your life as a pilot in a large airline. The seniority business causes a pain for the management but it does also create stability. Your seniority is based on your course date and as I'd been furloughed, because of the redundancy, I was on the first course. There were quite a few courses after me, so they would be below me in seniority. On my course, we had all joined on the same day, so age determined seniority. Those who were older than me, and most of them were, were senior to me, and I was senior to whoever was younger than me. Of those sixteen pilots, I was the last one to hang up my flying gloves. Dave later joined Cathay Pacific, having done half a dozen years with Britannia, and this was not uncommon. There were quite a few Britannia people who went to Cathay. They regarded our airline as a good feeder because we had good standards.

As I began this exciting new phase of my career, I told myself I had always known I had to work hard to succeed. If someone had asked 'How confident are you of being successful?', I definitely wouldn't have said: 'It'll be fine.' I knew the flying would be demanding and I was very concerned about the simulator, so I was very nervous. I had to sit a general exam for pressurisation as a separate module because it was one of the CAA modules I had never done. I also had to do AC electrics, which was a nightmare. I never understood a word of it. It was just trons and volts and squiggly amps – who cares? Star windings? What the hell does that mean?

We spent a few days covering safety procedures, which included opening doors with the slides, using fire extinguishers and smoke hoods, and getting into dinghies in the swimming pool to practise ditching scenarios. We started at the beginning, which, as always, was APU (auxiliary power unit). The main teacher was Dave Brown who, if he didn't get the f-word at least twice in every sentence, probably felt he wasn't trying hard enough. He was an old-school engineer and had been Jamie Alderman's flight engineer, so was guaranteed to look after me. One of our other captains, Hoof Proudfoot, later died in a WWII American Lightning twin-engine aeroplane while performing an aerobatic display at Duxford. He was a captain at Britannia and used to tell the story of flying a Spitfire towards Germany a few years before his death. He called up and asked if he could go direct somewhere, instead of following the route. The German controller replied: 'The last time you did not ask!'

The training was all talk and chalk, there were no computers. The APU would give us electrical power for the aircraft and also air to start the engines. Then we moved on to engines, hydraulics, pneumatics, air conditioning, flight controls, electrics, DC electrics, AC electrics, bus bars… on and on it went. Dave Brown knew everything about the aeroplane, and he kept us entertained with stories and by swearing a lot. In the dead part of the afternoon, if you looked like you were dozing off, he used to chuck wooden board rubbers and you were more than awake thereafter. I would go home at night to Olney, where I was staying, and spend the evening studying. The ground school went on for three weeks and by then we had completed our technical exams. We had also completed our SEP training - safety and emergency procedures.

A month later, it was time to go into the simulator. This was Britannia's own simulator at Luton. It was with a chap called Tommy Thompson, one of the training captains who were gods to me in those days. The first time I climbed into the sim, I was very nervous. I was in the right-hand seat for the first go and, because we were both co-pilots, we would swap seats after each exercise. We did the pre-flight procedures, started the engines and David, in the left-hand seat, taxied out for my take-off. I had thought it all through many times in my head but once we got airborne and retracted the flaps, it was like riding a tiger. It was oscillating and it was rolling, it was so light on the flying controls. You were supposed to fly it almost with your two fingers but I had both hands on the control column and, worse, I was squeezing in rudder at every turn, which I shouldn't have been doing.

I quickly learned that this jet was not an aeroplane I could fly in the manner I had been taught to fly. My biggest problem was over-controlling the aircraft. I was used to climbing the Trislander at 400 feet a minute and this thing was climbing at 3,000 feet a minute. Suddenly, you would look and think: 'I'm at 20,000 feet already.'

After three or four days, we started single-engine stuff. That day, it was my turn to take off and have an engine failure at V1, the 'go no-go' speed. The idea was to carry on the take-off on the remaining engine and climb away. We always calculated and used a wet V1, which was the decision speed for a wet runway. That meant you still had a bit of accelerating to do on the runway after the engine failure to achieve VR, which is rotate speed. At that point, you gently rotate the aircraft – tilting the nose up - at two degrees per second until approximately 10 degrees nose up and then climb away at the V2 speed. As the failure is introduced, the aircraft would swing towards the dead engine, I would squeeze in rudder to keep it running straight, and accelerate down the runway.

I was taught to lock my legs because the rudder was so powerful and any unnecessary rudder input would cause further problems in roll. I'd rotate a bit slower after an engine failure and then climb away, get the gear up, and try to keep it under some control. That was the plan. I climbed into the seat for my first go at a 'V1 cut' and off we went, accelerating down the runway. The engine failed at V1, as expected, and the aircraft started to swing. I stopped the swing and kept it reasonably straight. Then David called 'Rotate' and I gently pulled back to start the rotation. It lifted off and immediately started rolling. I started trying to control the roll - and we rolled on to our back and crashed. The screens went blank, the sim dropped off the jacks, and it was dead. We had to get out and call it a day. Did I do all that? Was it all down to me? Was it a sim problem? We never knew. It was our last detail with Tommy and he took us to the bar.

I went back to the house where I was staying, phoned home and said to Ruth: 'Don't sell the house.' My biggest concern was that I didn't know what I had done wrong. We came out of the simulator because it was busted and we all went home. I had a very strong feeling they might fire me, and even if they didn't, I might not make the grade.

The V1 cut, as it was called then, was the big thing to get right and I had just blown it and crashed. I still don't know why it happened. Was it too much rudder? Or simply the simulator being at fault, as they sometimes were? I had got to 100 feet and the thing just rolled.

We had finished with Tommy and I was getting a new instructor called Graham Freeman, who was an examiner and a completely different character. Tommy was a punchy, drinking bloke. He was ex-military and there was a marked difference in the way he instructed. Tommy just criticised: 'That's not very good, do it again.' Graham would say: 'This is what you do. You squeeze a little bit of rudder in and lock your feet.' I did the V1 cut again and it was much better. My second take-off was fine and away I went.

Taming the beast

I was not the only one who was nervous but our new instructor was reassuring. He would say: 'Look, guys, you're not coming out until you've passed.' We knew that wasn't necessarily true but he obviously had enough confidence in us to say that. As an examiner myself later on, I know you can say things like that but it can come back to bite you. Anyway, we did pass. In those days, simulators were not as true to real aeroplanes as they are now. As a result, the simulator was harder to fly. That's always the way they make them but, back then, it was considerably harder. In those days, there were no zero flight time-approved simulators so you had to do some training on the actual aeroplane. That is where things can be dangerous because we would be failing engines, or at least bringing them back to idle - these days, there is very little actual training in the aircraft and no single-engine work at all.

We took an empty jet to Seville for a few days. There, Graham took me, Dave Balman and another couple up and we did six take-offs and landings each. Then we swapped seats and the next candidate would have a go. What we needed was a long runway at an airport which wasn't busy and with a good weather factor. We also had to do some night circuits, which again were quite different to how things are now. Then came time for the single-engine work and when he failed it, it was hardly noticeable because of the momentum of the aeroplane. It was our first taste of night stops in the company and I found it all very exciting. Only one of the pilots on the whole course failed to come to Spain and that was because he hadn't done well in the simulator. That chap was one of the RAF fast jet pilots. I was amazed - he had a long background flying Meteors. He eventually got through the course but as the years went by, he was never promoted. The rest of us were all signed off.

I was putting a lot of work into the different scans, the checklist responses and the general order of the pre-flight preparation. At Britannia, if the call was a thousand below, that's what you said. If you said 'One to go', you'd get a bollocking. They wanted it all done by the book, so any two pilots could fly together and know exactly what the other was doing and what to expect. Britannia's training department had an intimidating reputation, especially in the early years, but it made me work and study hard. There was pride in knowing that if Britannia was happy with my performance, I must be doing okay. I was used to flying around Scotland and England. Britannia, on the other hand, always went abroad. That was a whole new world to me: not just flying at high altitude and speeds of eight miles a minute, but I was worried about the RT and foreign controllers. Would I understand them?

I started my line training with my first training captain and I worked out fairly quickly that he was not a very nice man, self-opinionated and pompous. He was a line trainer, the lowest position for training captains, and he was an awkward fellow. We also had Chris Blake, who was going to be in the third seat because, at the very beginning, the captain may need some help from a fully qualified co-pilot, so they carried a qualified first officer. Chris was with us for the first six flights. The first I did was to Zurich, a busy, complicated airport. It was quite a short flight, so there was not much time in the cruise to think about and plan the descent and approach. We all had our own set of Jeppesen Charts, so I could study them at home. And boy, I did! The idea was that you flew three miles for every 1,000 feet you descended - we were basically gliding with the engines idle. So if you were at 33,000 feet, you would require 99 track miles. There were also variable corrections for head or tail winds and aircraft weight, plus a bit to slow down. So you would request descent at, say, 112 nautical miles to touchdown and hope ATC would let you down.

I'd never done anything like that before. Flying non-pressurised aircraft, you have to come down at a low rate of descent to protect everyone's ears, but this jet could come down at any rate. My first flight was an hour and thirty minutes, although it felt like five minutes. We had a calculation that, at 10,000 feet, you wanted to be forty miles from touchdown and at 250 knots indicated speed. If you were higher than that, or faster, you would need the speed brake and if that wasn't enough, you would need more track miles to touch down – or a miracle.

There was a general agreement that you didn't go faster than 250 knots below 10,000 feet, unless you asked ATC. At this stage, I certainly wouldn't want to. A speed of 250 knots was still four miles a minute, possibly five with a tail wind. That was what I would plan for. I would have 'gates' that I would be aiming for, and I would say: 'I feel a bit high.' The trainer would say: 'Yes, you're about 2,000 feet high on the profile.' There was nothing visually to tell me, unlike in a B757 or a B767 later on, and the training captain wasn't very forthcoming at all. On my third flight, I went to Monastir in Tunisia, a big destination for us until they started having civil wars there. From a pilot's point of view, that was a better experience because it took two hours, forty-five minutes going, and three hours, twenty minutes on the way back. That gave me time to settle in, because I had to do the take-off, top of climb and navigation checks, and I would be working the radio and maps, and then thinking about my approach, which I had to plan and brief.

After those six sectors, the first officer got taken away, which was good - if it was not going well, he would have stayed. I did a couple more with him to Munich, which would have been RAF trooping flights. We used to do the trooping flights to Germany with all the RAF and Army people. We also did more flights to Monastir, Zaragoza in northern Spain, and Gutersloh, an RAF base in Germany. These are not very nice places to send a new pilot. The flights are only an hour and five, and with a flight like that, you hardly cruise. It's very demanding.

My next trainer, Bob Stockholm, with whom I did another nine flights, was a much better trainer, but he still took no prisoners. One time, I climbed out of Gatwick and he asked me: 'Where are we?' I went: 'Er… south of England?' I hadn't much of a clue. With Bob, I went to Palma, Faro and Monastir again, but out of Gatwick and East Midlands. I was sent to all the bases around the UK that we flew from. Almost all the pilots were based at Luton and we would go off for four or five days and fly from different UK airports.

The B737 was an animal, a monster that needs to be tamed, and at that stage I hadn't tamed it. I still couldn't land it very well. I was hitting the ground quite hard. When I got to my final line check, Dusseldorf to Luton, I passed it and was released to the line - but I was not allowed to do the landing unless I was with a training captain, which was a bit of a fudge. I couldn't land the jet, but I was licensed if something happened to the captain. I flew with a chap called Mike Harvey, who was a gentleman, and when I confessed my difficulty, he said kindly: 'That's unusual, Scott. Well, you do the take-off and fly, and I'll take over at about 1,000 feet and land.' We did that from Palma back into Cardiff and I didn't do any landings. A day or two later, I flew with Colin Sharples, who was a training captain and an examiner, to Hanover and back again, a trooping flight. I landed it into Luton and, as with all landings into Luton, it would not have been particularly easy because the damn place is on a hill. The winds swirl around a lot and the runway has a hump in the middle, so when you land at one end, you can't actually see the other end. In my time I have hit that runway hard, but it still has its famous hump - I should have flattened it by now. During my line training I had been landing the aeroplane, but not always successfully. I was missing the height of the flare, because there was no accurate RADALT (radio altimeter). I was having to start the flare visually at about thirty feet and then maybe squeeze off a bit of drift due to cross wind and take the power off. I would think 'This is going to be a bit rough', but it ended up a perfect touchdown. Next time, I would think 'This is okay', and it would go crunch. It's a bit like getting in a car and reversing round a corner. You know how to do it but sometimes it's perfect and other times it's a mess.

The landing of a jet is always a bit of an art. There is no one who can 'grease it on' every time – and, of course, 'greasing it on' is actually not the correct way to land a jet. The aeroplane has to touch down firmly enough for all the sensors to register contact on the ground so you can get ground spoilers and reverse thrust. If the aircraft lands smoothly, the speed brake doesn't automatically deploy. That is not a problem as long as the runway is long enough, but pilots know passengers like a smooth landing so they will generally try to achieve that if they have lots of room on the runway. You can't 'grease it on' at places like Luton or Bristol, but you can at Palma, which has a long runway. Later, if I was checking a crew and they 'greased it on' inappropriately, they would be in for a bollocking. The passengers might be going 'Oh great', but the poor chap at the controls would be getting a hard time from me.

Colin obviously thought my landings had improved and he signed me off for landing on the line. I then flew on my first line flight with a captain who happened to be a trainer and was another miserable so-and-so. We flew to Alicante, a place I had never been before, and I was briefing from the charts because that was all I had known - we had never done visual approaches. He let me go through all of it and then asked: 'Aren't we going to do a visual?' A visual? I had never been to Alicante. Damn! We were about to start descent and all my planning had gone out of the window. I had the feeling he was going to watch me dig a hole for myself and then give me yet another bollocking.

We had been taught visual circuits in the simulator but not a visual approach from 10,000 feet. I hadn't even considered a visual approach on this occasion or that it would be a normal operation on the line. Nowadays, pilots take the visual option whenever possible, because it is usually more expeditious than the radar or procedural approach. On modern jets like the B757/767, there is a Magenta Line to follow on the screen, so if ATC say that instead of going from A to B to C, they are clear to go direct to C, they just go direct to C in the computer – and the Magenta Line goes direct to C. Easy. So pilots always know where they are: they have the map with the Magenta Line and can see the visual picture of where they are.

What we used to say in the B737 was: 'We don't quite know where we are – but we know where we are going.' Nowadays, pilots know where they are but they don't necessarily know where they are going. That is because they trust and follow the Magenta Line. Because the jets are quieter and a lot more automated, and because they're nice and cosy, people forget what they are flying. I always felt I would like to take new pilots to the end of the runway with a jet landing over them at fifty feet, so they can actually see this howling piece of metal hurtling towards the ground.

There was a story that Colin Budenberg and I used to tell training captains at training conferences to demonstrate a lack of situational awareness and complacency. One of our training captains was flying a B757 into Paphos in Cyprus and training a new second officer. The weather was nice so they had briefed that the second officer would carry out a visual approach by breaking right into a left-hand circuit and flying downwind. The correct procedure to fly downwind, according to the Flight Crew Training Manual, is to fly downwind at 1,500 feet, at 170 knots with flap 5 in a B757, which is about a third of the flap. As they pass the end of the runway, abeam the touchdown point, start the stopwatch corrected for wind, put the gear down and select flap twenty, and run the landing checks. Then at 45 seconds, give or take wind, start turning in and take the rest of the flaps and establish on a three-degree descent. That should put them on to the correct profile for landing.

This chap was downwind, hand-flying at 1,500 feet, which is not very high, especially at 170 knots. The training captain was sitting on the left-hand side, with the runway down on his left. The second officer was pilot flying and he was running downwind. The idea was to be about two miles out visually at 1,500 feet. They had achieved that and the captain looked over to the second officer, who was hand-flying the aeroplane – the autopilot was not on. When I say 'hand-flying' it, he was not actually holding on. He was leaning down at the side of his seat, fiddling in his briefcase. He came up with a manual navigation wheel in his hand… and the captain, not unnaturally, said: 'What the hell are you doing?' He replied: 'I'm just trying to get the reciprocal heading because I'm not very good at that.' The captain yelled: 'There's the runway down there! You don't need the reciprocal, you need to fly the bloody jet.' This young chap had such a lack of situational awareness, and such a lack of fear of this screaming lump of metal, that he let go at 1,500 feet. Unbelievable.

Culture shock

Inga was coming up to two and Ceinwen, our new baby, was just five months old. We bought a house in Olney, and that summer saw my first operational flight with a normal line captain, a nice gentleman called Don Kirkpatrick. We did a couple of flights out of Newcastle. Usually, you would fly 'leg and leg about', and the captain would say: 'Which way do you want to fly?' Occasionally, he would say: 'I haven't done much flying, I'll fly it both ways.' If the weather was bad, the captain would fly that leg. Britannia was still very much a captain's airline in those days. Some of them hardly even let you look at the take-off performance manual. You were 'the boy', and I was very comfortable with that. If someone said 'Do you want to be a captain?', I'd have said: 'No way!' I had three stripes and the old uniform was very naval, with a double-breasted blazer, double rows of buttons and a white hat. We weren't working for a hick operation. In years to come, when I went to Virgin, neighbours would say: 'Oh, you're working for a proper airline now.' I would think: 'You have no idea!'

Next time, I flew with a chap called Ian White and said: 'I'm still struggling with my landings.' I plonked it down so hard in Ibiza that some of the oxygen masks in the cabin fell down. He said: 'Bloody hell, you're right – you can't land the jet!' The crew didn't take prisoners. They came in after the passengers were off, asking 'Whose landing was that?', and the captain pointed at me. They left muttering about L-plates, boy pilots and back pain. In those days, we hardly ever flew out of Luton. A fleet of Mercedes Hallmark cars took the crews off to various airports or hotels. We used to go away for about six days, and that was our life. We were away a lot. It quickly became my habit to look at my roster and work out if the captain I would be with was a good guy, and was I going to learn something from him? I was beginning to learn the off-duty culture at Britannia. There were no male stewards in the airline. Cabin crews were all women, and the pilots were all men. The one exception was one female co-pilot who married Graham Freeman, the training captain who passed me in the simulator check and a lovely man. A lot of the pilots had been married multiple times and it didn't take long, even for someone as naïve as me, to figure out why. The joke with the pilots was: 'I've been married for twenty-five years – to four different women.'

The culture was that the pilots would have an after-landing drink. As you taxied in, the girls would bring us in a whisky or whatever your tipple was – while we were still taxiing the aeroplane. Today, you would be shot for that. In recent years, one of our captains was breath-tested, although it was eventually proven that he was okay. Nevertheless, he was grounded for a month because he had to have blood tests. After the results were known, there was an interview with management. He admitted drinking, but claimed in mitigation: 'I had a couple of beers last night with dinner.' In the old days, the golden rule was always 'eight hours from bottle to throttle'. That didn't mean it was eight hours from a drink - it was eight hours from a good drink to starting the engines. It was actually frowned on if you didn't go out as a crew. The girls talked of the pilots as 'the men' and we called the cabin crew 'the girls'. Nowadays, that might not be acceptable, or even accurate.

A few months in, I looked at my roster and I had four clear days in Tenerife. They put us all in a hotel with an allowance - a couple of quid an hour every hour we were away. I couldn't believe that someone would send me to Tenerife and leave me there. One captain, Rick Morris, was a nice guy but did everything by the book. He was ex-RAF, and I travelled in the back of a car with him to the hotel in Manchester. Our flight the next day was Manchester to Heraklion, just before this Tenerife trip. I had gone on line at the beginning of that summer of 1982, and this was August, so I had had a fair amount of flying hours, but every time I flew, it was with a different captain. Rick asked how I was getting on, and I admitted I was struggling with the descent.

By now, I had got the landing sorted, but I found that, once I had started the descent, it was difficult to judge if I was high or low on the descent profile. I had track miles written down on all my charts, which I updated weekly. Things like: 'From touchdown to this beacon six miles, from this beacon to this beacon...' I had the right idea but I was still struggling. He admitted he had really struggled, and he also told me what he had done to sort out the various profiles. Like me, it turned out that he wasn't very good at maths. I could see other pilots doing complicated mental maths but I couldn't do that and fly the aircraft. Rick said: 'All you need is your three-times table. You maybe need your seven or eight-times table, for miles a minute during the cruise, but your three-times table will sort out your descent problems.' He added, and it's something I have always remembered: 'Don't make it too complicated. It doesn't need to be to the nearest 100 feet. If you look at your altimeter at about 22,500 feet, wait until it gets to 22,000 feet, then multiply by three and add 10 - that is what your track miles need to be. So 66 plus 10, therefore 76 track miles are required to touchdown. If you can't work out roughly where you are, don't panic, leave it a bit and then do it when you know. If you really don't know where you are, ask ATC for track miles to touchdown.' You can't always get track miles from the Greeks or the Spaniards, but in the UK the radar controllers are very good and it's quite all right to ask: 'What's my track miles?' They'll give the answer.

By the time I got out of that car in Manchester, things had been clarified enormously. He had given me some simple tips, none of which were mind-blowingly difficult but they were about to make things considerably easier for me. I knew about three times the height, but he said: 'Use the KISS principle – keep it simple, stupid!' This simple advice lifted a veil and it made a huge difference.

The big authority gradient between captain and co-pilot in the post-Second World War period and up to the early 1980s has since reduced considerably. As a co-pilot, I witnessed captains barking orders about what they wanted. The co-pilot was the boy in the team with little to offer the captain. Statistically, it became clear that aircraft were crashing not due to a lack of pilot knowledge or handling, but to a breakdown within the flight crew. The co-pilot had the knowledge but wouldn't say because he was scared of offending the captain. On one occasion the flight recorder showed that the co-pilot's situation awareness was correct but he was scared to tell the captain and they hit the hill. After years of CRM (crew resource management) training, the authority gradient pendulum might have swung too much the other way. Captains are sometimes reluctant to take control when necessary for fear of hurting the co-pilot's feelings. Judgment, as always, is the key, and understanding the fine line between being autocratic and letting the other pilot run the show is crucial.

Many a flight safety report summary has pointed out that the captain should have taken control. Feedback was: 'You should have taken command.' For example, a first officer might be flying the approach in poor weather with strong cross winds and if you could freeze the aircraft in the air, as you can in the simulator, you could ask the captain: 'How do you feel?' He might say: 'I wish I was flying the aircraft.' If you could ask the first officer the same question, he might say: 'I wish the captain was flying the aircraft.' The first officer would be flying the jet because the captain doesn't want to hurt the first officer's feelings. Now it's made clear: 'It's not about feelings – it's about not hurting the bloody jet. Feelings you can sort out in the bar!' Captains need to feel comfortable saying: 'I will fly this approach because the surface wind is all over the place.'

Later in my career, flying a B757 and climbing up to Flight Level 190 (19,000 feet) out of Birmingham with my first officer as pilot flying, London ATC suddenly called us and said: 'Thomson 9DW (9 Delta Whisky) stop the climb FL080. AVOIDING ACTION, TURN RIGHT HDG 215deg!' We were already higher, passing FL083. The first officer started to select FL080 into the altitude window but we were still climbing. ATC immediately said: 'Thomson 9DW. AVOIDING ACTION. TURN FURTHER RIGHT 225deg!' I told the first officer to take the autopilot out but nothing was happening, so I called 'I have control' and disconnected the autopilot. I rolled on bank and stopped us climbing but, by this time, I had warning calls of 'BANK ANGLE, BANK ANGLE' as we were rolling past 40deg, plus TCAS (traffic collision avoidance system) was shouting 'Traffic, Traffic' as well. At that point, I saw an aircraft coming towards me and turning to its right. I asked ATC to confirm I was clear to climb to FL190 and return to my route. The ATC controller sounded mortified. I guess he had seen the two radar traces merge and thought: 'I have just killed 300 people!' I felt for him. The point is that there wasn't time for me to think about my first officer's feelings. I had to fly the aircraft out of trouble. On landing, we had a good chat and dealt with the emotions that needed to be talked about.

When I joined the airline, a training captain would earn more than one of the directors. The boss was a captain and the chief pilot was also on the board. A line captain earned more than most of the senior management. As a result, most of those captains wouldn't take a lecture from anyone except a senior trainer or pilot manager. If a non-pilot manager started criticising a captain, the chances are he would have simply said: 'Who do you think you're speaking to?' He would have just walked away. If a captain did that now, there would be a bollocking.

Some of the changes were needed because the pilot directors really didn't have the business knowledge. On the other hand, they did run a pretty tight jet operation. Conversely, the new directors soon found out how many people worked for the airline but did not concentrate so much on running a tight flying operation. Somewhere in the middle was the right answer.

Breaking in the new boy

In my first year at Britannia, I did eight simulator details of four hours to get rated on the B737. After a gap flying the line, I went back into the simulator for CAT 11 (category two) training. This was somewhere between autoland and a normal instrument approach, where the autopilot would be engaged until 80 feet. On the B737 you still had to do a manual landing, but in poorer weather conditions than the standard 550 metres visibility below 200 feet. A CAT 11 was always a captain's landing but we needed to be trained in the specific roles of the CAT 11 operation.

Then I went in again for my first recurrent base check, as it was called then, which was two days of three hours. I worked really hard and was a bit deflated to hear at the debriefing at the end of day one: 'You're okay but you're a bit behind the aeroplane.' I went to the loo and the other captain I was with, who was becoming a training captain, came in and could see I was unhappy. He said: 'I thought it was fine, Scott. It wasn't sparkling, but it wasn't bad. I've seen a lot worse.' That lifted me and made a big difference to my confidence.

I'd only been on line a few months and in that time you don't experience all these emergencies during normal operation because I was flying a serviceable aeroplane, by and large, from A to B. I'd spent weeks studying but hadn't got on top of everything because I'd been reading general stuff about flight time limitations, rules and regulations. I learned that all that had nothing to do with a simulator check. The night before I was going in the sim, I needed to be thinking about engine failures and clean-up profiles – the actual stuff I would need. What I would do today is 'fly the armchair'. I would sit in my armchair and think: 'Right, take-off. On the runway, power up, keep straight, elevator control forward until 80 knots, and then centre the control. Engine failure – squeeze in the rudder then freeze my leg, pause, keep the aircraft straight, don't try to get back to the centre of the runway, just keep it straight down the runway, rotate, gentle rotation, positive rate of climb, gear up, feet locked, check the controls, check we're climbing, increase power on the remaining engine if needed, small smooth control inputs, keep it stable.' The other pilot in the sim will put out a Mayday and at about 400 feet, I will ask for the lower screen to identify the problem and then decide: 'Do we have any memory action?' I'll go through all that sitting at home, getting it clear in my mind. Then I will do the same for the single-engine approach and go-around. It is what we do, to a lesser extent, before every take-off. I've always admitted to everyone that I am not a quick learner, and this was a whole new experience. When you're in the simulator, you don't know what's going to happen. I was totally overwhelmed by the aircraft, as most people are. Years later, of course, I could look back and realise I was learning things all the time and slowly managing the aircraft, as well as my stress level, so as not to be overwhelmed.

Then LOFT – line orientated flight training – was introduced in the simulator and the idea of LOFT was that it was on the simulator but in real time and on a normal operation with passengers aboard. During the flight, you would deal with whatever failures the trainer gave you. You only had four hours so you might start on the ground but finish mid-flight, or start in the cruise and take it from there. The idea was that we were flying from, say, Palma to Glasgow, and at some point we would have an engine problem and divert into Barcelona. I would pretend to do a public address, and deal with cabin crew and engineering – the aim was to build crew co-ordination. It was a new concept, which we've since expanded.

The instrument rating was hand-flown so I flew the jet like a Cherokee. I could use the flight director, although that was pretty poor in the old B737, but no autopilot. They were watching to see if I was accurate. I was flying within tolerances and I would go into a hold, check my entry into the hold, and then shoot an approach, go round and land. That was the classic instrument rating test and it was almost single crew. The other pilot there would bring the gear up but all the flying was done by me. Later, this was changed, when the CAA acknowledged the obvious facts that it was a two-crew aircraft with multi autopilots, so why not use them?

In those days, the captain would have to do something fairly wrong to get a bollocking. The first officer was there really to take the blame.

That first time with the captain who was becoming a trainer, when we did the go-around he brought the gear up before the flap. You would do that in a little aeroplane but not in a jet, and it caused all sorts of warning wailing noises. The debriefing afterwards was: 'Just be careful.' If it had been me, I would still be there getting bollocked! I was checked every six months, which was normal, and I started to get on top of it. I had learnt all the SOPs (standard operating procedures) and the non-normal procedures. I had studied the QRH (quick reference handbook) and spent many hours sitting in an armchair thinking through all the various failures I could come across. My preparation became more finely tuned, which made a difference.

Later that year, I flew with the trainer I had first flown with, who was a miserable so and so, but I had got used to him and managed him by asking questions. I found with the older pilots, I would just say: 'Are you ex-military?' They'd say yes, and I'd ask: 'What did you fly?' Quite often, they would say: 'The Vulcan V-bomber or Meteors.' Not many were ex-war, although some were, but I was still interested. They would talk about their time in the RAF or Navy, which would break the ice. I was in my late twenties and they were late fifties. I could have been younger than their kids, and in those days that represented an even bigger generation gap than it does today. They would talk, for example, about being stationed in Singapore, and it was all very interesting. It was an important skill for me, because I would be away and having dinner with people I wouldn't necessarily choose to have a drink with in a bar. If we were night-flying, quite often the captain would say: 'I'll see you at 12 o'clock for a beer, help you sleep in the afternoon.' I couldn't do that. If I had a beer, I would be destroyed for the night flight, but they would have a pint and I would still meet them for a chat. I was learning a new way of life and new social skills.

We were usually away with the cabin crew. Over the next couple of years, bases opened in Glasgow, Newcastle, Manchester, Gatwick, Cardiff, Bristol, Birmingham and Leeds. They all had their own cabin crew and eventually the pilots were allowed to move to places like Glasgow, Manchester, Gatwick and Bristol. About ten years after I got my command, the job changed because if I went to Manchester for a few days, I would often fly with a local first officer and cabin crew, so I was on my own – though in the hotel there would no doubt be a bunch of crews doing other flights, so you were seldom completely on your tod.

Towards the end of 1982, I met a captain I got to know pretty well and had great fun with. Chris Jackson would have been in his late forties and to me, at twenty-odd, that was still quite old, but he was wonderful. He was a training captain but fun to be with, and he used to have great fun with the girls. He was a bundle of laughs. I would look at my roster and think: 'Oh good, I'm flying with him.' We were flying somewhere at night and were crying with laughter so loud at the front that one of the girls in the cabin called through: 'The passengers can hear you!' That just made us howl even more. I was at the point of being completely out of control, crying with laughter. At that very moment, French ATC was trying to call us – 'Britannia 322 Alpha, this is France' – and we were just howling. Nowadays you'd look out of your window in a situation like that and find a fighter jet beside you.

There were a lot more characters around in those days, although some weren't as much fun as others. The odd one or two old captains insisted on flying the jet their way, which was frowned upon. Sometimes I would sit there as we would be in a dive at 330 knots and think: 'I have no idea what you are doing and no idea where we are on the profile.' I didn't have enough experience to think: 'We're actually close in but we are diving the height off.' They would just say: 'Oh, this is the most efficient way to do it.' It made it impossible to monitor the approach. There have been times when I have been overhead the airfield at 33,000 feet because ATC hasn't given us descent clearance, and you learn that diving the aircraft at high speed causes lots of drag and gets you down very quickly. But below 10,000 feet, high speed isn't the best way. The speed brake is more effective at high speed so it's a balance. In those days we didn't have any vertical profile so we were always just doing the maths thinking: 'Right, where am I?'

It was in the winter of 1982-3 that I fell out with a captain in a big way. I flew with this chap who was very military and not very friendly. We were doing a flight to Dublin with newspapers, which we didn't often do. I was called from standby for a night flight and the weather was snowy. In Dublin, they were unloading the aeroplane and air traffic called up with a message to call Luton, because in those days there were no mobile phones and we didn't have an HF radio. The captain jumped in a car and off he went. He came back and didn't say a thing. We started engines and departed for Luton. In the climb over the Irish Sea, air traffic called and said: 'Just to let you know that at Luton there is a blizzard.' The captain said: 'Okay.' I asked him what Luton said and he replied: 'If you can't get into Luton, just go to Manchester.' I asked: 'Are we diverting to Manchester?' They were happy with the aeroplane at Manchester. 'No. We'll go to Luton.' 'Why are we wasting fuel if the company can use the aircraft at Manchester and the weather is good there?' We got overhead Luton and were holding at about 20,000 feet, and the braking action, which is measured at three places on the runway, was poor, poor, poor and they were trying to clear the snow. Nothing was happening - one of the snow ploughs had broken down. So I said: 'Why don't we go to Manchester? We are below minimum fuel reserves for Manchester and the whole of the south of the country is under a blizzard.' He said: 'No, we're all right. We'll shoot an approach.' I said: 'But the braking action is poor for the whole length. If we make an approach and then divert, we will be on fumes.' We had this disagreement and I said I wasn't happy, but it wasn't acknowledged. I could just see myself off the end of the runway in the field with the managing director looking through the window saying: 'I pay you good money to stop idiots like that doing this to the jet.' That was my job. I said to him: 'I'm now going off air and I'm going to call Operations on the other radio and I want to speak to a manager.' There was a huge silence and he said: 'Okay, I disagree, but we will go to Manchester.' We went straight to Manchester, where we landed with not a lot of fuel because we had been messing around over Luton. He was fixated with getting into Luton. When we got off the aeroplane, he said: 'We'd have got into Luton fine.' He was just a pretty unpleasant bloke, and I hadn't realised that he was known for it because I was too new.

Those kinds of situations are now taught in the sim and in the classroom. Some years later, when I was running the crew resource management (CRM) programme, something happened to that captain on board his aircraft and it should have diverted. He collapsed and the first officer started to divert, but then he recovered, came back into the flight deck, took command again and carried on to destination. It was then made clear that once one of the pilots was incapacitated, the aircraft must divert and land as soon as possible. The flight could not continue, even though the person might then be feeling fine. At the time I was a union rep and because of that, and because I was running the CRM side of things, I was asked to debrief him. I phoned this chap up, and he was absolutely charming and completely open. He told me what had happened. He didn't like going to the toilets on the aeroplane and had held it in so much that he had collapsed. He was contrite and said he had learned his lesson. After I had been made captain, I heard stories from other pilots who had flown with him. On one flight, he and the first officer had a screaming match before take-off and the cabin crew and passengers could hear it all. The first officer screamed: 'Right, that's it. I'm getting off!' Clearly, by the time of my interview with him, he had mellowed very considerably or was on his best behaviour.

It's hoolie time!

Lots of the captains had wives who were crew members. Liaisons between the pilots and the cabin crew were completely the norm. So much so, that when I talked to people at work about Ruth, they would say: 'Oh, was she crew?' Pilots were always going away and they would meet female crew members and others in a social atmosphere.

We were all staying in nice hotels and in those days we would drink after a flight, no matter what the local time. If we arrived at six in the morning, we would go and have a drink. This was particularly true of people like one lovely chap who was a New Zealander. He didn't eat much at all, just drank little beers. We would be having dinner and he'd just smoke all the time and drink, but he was wonderful fun. He was also a really good pilot and aircraft handler. When things were getting serious as we descended down through 300 or 400 feet, he would disappear down the side of his seat and into his bag - and come up with a lit cigarette. He was always a bit irreverent: the company eventually told him not to wear his Mickey Mouse watch. I used to see him on the roster and think: 'We're going to have so much fun.'

It was always made clear by the company – and this was also mentioned during the command course – that we should never leave a mess behind us in the hotel. On one occasion, going into Bristol one night with him, we had descended really low for some reason. We were not illegally low, but we were over the city of Bristol at about four in the morning. Radar had put us there. On top of that, we had the flaps out, the gear was down and we had to level off. We had to put on a huge amount of power just to fly level, so there was a lot of noise. We might have been at 3,000 feet but the old noisy B737-200 with a lot of power on at that height wakes a lot of people up. My colleague had absolutely no remorse. He looked at me and said: 'Good morning, Bristol!' We could imagine all the lights going on below us and thousands of people moaning. Again, it left me howling with laughter.

We landed and all went off to this hotel where we had a room party. Dawn was coming up and we were all getting a bit tiddly. The Number One wanted a dance, so a couple of the girls took the bed and put it up against the wall. In doing that, they took out a wall light. The captain phoned down for the manager, gave him a beer and he joined in the fun for a bit. We were due to go flying that night, so the manager took us to the kitchens and we all made breakfast with him. Our captain had obviously said to him 'What do we owe you?', which is what the company expected.

The chaps who flew the jets in the early years were from the war and that was what they did when they came back from a raid. They would go and have breakfast and then a good drink so that the next night they could go out in their bombers and do it all again. That was what the drinking culture was still based on. To start with, at our relatively young airline, there were only girls in the cabin crew but then stewards started to arrive. We used to have great games. If there was a new girl on the crew, we would play a game where we would sit on the bed and pretend to be different parts of a Wellington bomber. One pilot would be at the bottom of the bed in the middle in the cockpit. One girl at the top of the bed would be the rear gunner firing her machine gun and a girl either side would pretend to be the aeroplane engines, making appropriate engine noises. Then someone would go and get a bucket of ice and water and the captain would shout: 'Engine fire number 2!' And the water would be thrown over the new girl who was acting the part of the number two engine. A lot of senior management I knew had been in various scrapes in their time. One of them was always called 'the man in the hat rack' by Ruth because he got so drunk on one occasion that we actually put him in a hat rack over the seats in a train. The training department had arranged a dinner and dance on board a steam train for all the trainers and their partners. You see the same kind of thing in The Dam Busters film, where they're breaking up the mess and having a complete hoolie.

When I was captain of a B767 and we were in Orlando, we had gone down to Disneyland to Pleasure Island where the adults go. We had a minivan and I said I would drive, so I wasn't drinking. As I was driving back, the crew were singing 'Scottie, we love you' and were all kissing my neck. At one point, I thought: 'That feels prickly, that's a steward!' We got back to the hotel and we weren't flying the next day, so the crew were going to have a room party, a hoolie. I could see they were all off on one and decided I was going to go to bed. I was sober and I was past all of that. I pulled the first officer out and said: 'Look, son, I'm going to bed and if you're going to this thing, you are in charge and you've got to take control. I can't get into trouble as captain if the hotel's broken up and I'm in bed. All eyes will be on you.' Next morning, I said: 'What happened?' He replied: 'They were up on the 8th floor and we were all having a drink. The steward got naked and he was outside the window on a window ledge, stark bollock naked with shaving foam round his private parts, with a sheer drop below and no balcony. I went to bed at that point...'

In Japan one time, I had to buy a complete new bed after it was damaged, and that was very expensive. It got burnt and, because I was co-pilot, I had to go out and buy the damn thing. It wasn't unusual to climb in the transport to go to work and for the captain to say: 'You all owe me a tenner.'

In those days, we used to carry cash. Every trip, I'd carry a float for landing fees. US dollars, usually, but sometimes I had pounds as well, depending on where we were. Captains could write money off pretty easily on 'ground handling'. I was standing under a jet in Kingston, Jamaica, after diverting and was told: 'You'll never get out now, the airport shuts.' I took a fistful of dollars and said 'This is yours if I get out' – and I did. In that instance, I couldn't write down 'a hundred dollars given because they wanted a bribe', so I put down 'ground handling'. The company was slightly naive because all the pilots had to watch a company presentation about bribery and corruption, but sometimes it's the only way to get off the ground. Going down to Mombasa one time, the aircraft in front of me said there was an ATC strike there, and I relayed the message to other aeroplanes behind me. This would mean holding for an hour and I calculated that I hadn't enough fuel for that. When I got there, I was told to take up the hold, and I didn't really have the wit to say to someone: 'Why don't we just buy the controller a car? Whatever it takes.' I diverted to Dar es Salaam, which might as well have been the moon. No one in Britannia had been to Dar es Salaam - and I arrived there with 270 passengers.

I had a company phone – this was late 90s – so I called up and they said: 'Can you get in and out in 20 minutes?' I said: 'I've been here for 20 minutes already and no one's even approached me. I've got to find an agent. We've got to get fuel. I'm here for the day.' They replied: 'Okay, best of luck.' They phoned back and said: 'We're sending down a couple of Thomson reps from Kilimanjaro. They'll be there in a few hours.' I found a handling agent at the airport who could sort out steps and fuel. We needed to get some bags off the aeroplane but the B767 is not bulk loaded - all the bags are in containers and you need a high loader to get them on and off. We also needed to clean the aeroplane and, of course, we needed fuel. It wasn't a long flight back but I wanted to arrive there so I could hold for a couple of hours if required. And we needed somewhere for everyone to stay. Fortunately, a new Intercontinental hotel had opened and they had lots of rooms, but they didn't have enough coaches - we ended up using pick-up trucks with passengers sitting in the back. We shut the jet up, which took hours. I said to the crew: 'Will you clean the aeroplane, tidy it up and I'll buy you all dinner?'

One particular passenger was already kicking off at this stage. He had had enough, and was saying he needed to get to Mombasa. He wanted me to hire him a private jet. 'You're on a chartered flight to Mombasa on a holiday and you expect me to get you a private jet? That's not going to happen.' To add to my ever-worsening headache, the authorities wanted 100 dollars per passenger for tax to get into the country. I offered 20 in cash and they went: 'Done.' I thought: 'Welcome to Africa!'

I did a deal with the authorities and they wanted $1,000 for the landing fee. I just about had that on me. The hotel was wonderful. I said: 'Feed the passengers but don't give them alcohol.' They even had notices in the lifts saying 'Britannia passengers, food available' and what time we were leaving. By then, the reps had come across from Kilimanjaro, but it became clear they were only interested in making money out of the situation. I phoned Luton and said: 'Recall them, they're not helping me at all.'

Before I was properly rested, I went down to the lobby where some of the passengers were mingling. The private jet character came up to me, complaining again, and I lost my temper with him. I immediately regretted it and said: 'I'm sorry about that.' I wasn't sorry that I had told him to bugger off, but I was sorry that I raised my voice. He had already asked for a private jet. What did he want now, a space shuttle? I am a pilot but I had organised everything in this crisis, which is exactly what we do in situations like that. If I had been working for Virgin or BA, there would have been someone there to do that for me. In that hotel lobby in Dar es Salaam, I was furious. I was at 10 on my scale of 1 to 10, and it was the only time I ever shouted at a passenger through anger and frustration. The outcome was still: 'No, you can't have a helicopter or a private jet.' I walked away.

There has been at least one occasion when I've bollocked a whole cabin full of passengers. I told them: 'We're not going anywhere and I'm going back into the flight deck for my coffee.' They were all whingeing because we were hours late. I stood in front of them, which I didn't have to do, and explained that I wasn't getting airborne with them: 'You're all cross and as soon as we're airborne, you'll take it out on my crew. When the Number One tells me you've calmed down, we might go back to Glasgow.' A wee while later I was told: 'They're okay now, Scott. They've calmed down.'

I interceded quite a lot where some captains would leave it to the crew. The way I saw it is that passengers will pick on the crew but they won't pick on the captain 99.9% of the time. So I was known for looking after my crew. There were many times when I wandered around the B767 down one aisle, maybe if we were on the ground with a delay. The passengers might be eating breakfast because I had said to feed them on the ground, and I would go to every seven or eight rows and say: 'How's the food? Is it edible?' They would say yes, and I would reply: 'Well, that's a first.' I would work my way around, answering questions, and usually when I got back to the flight deck via the other aisle I would have left them much happier.

When I was flying for Virgin, there was a huge delay in Gatwick once and the B747 captain didn't do a PA. I asked if he wanted me to do one. 'No.' We weren't going anywhere, so I told him I was going to the loo and went down the Jumbo aisles, where the crew were having a really hard time. Everyone could see the snow outside, but no official was saying anything. Anyone would have been hacked off, and if the captain doesn't do it in a situation like that, who was going to tell them?

At the Dar es Salaam hotel, all the transport arrived again. Some of the passengers were at this stage thinking this was a real adventure and were saying things like: 'You couldn't pay for this. We've managed to visit another country.' We got back to the jet and I went to pay my 1,000 dollars. Lo and behold, it had become 1,500 dollars. I said: 'But you said it was 1,000 dollars.' They said it was dark now so we were going to need the runway lights on. I only had 1,000 dollars and there was no bank at the airport.

I went back to the jet and found all the cash I could, about another 100 quid, the last of my ship's funds, and told the cabin crew to fill as many bags as I could carry with bottles of whisky, vodka and gin. I went across to the building with two carrier bags in each hand. I counted out 1,000 dollars, counted out 100 quid, and put these bottles very slowly, one at a time, in a big row of booze on the counter. Then I said: 'That's as good as it gets.' The guy behind the counter smiled broadly and said: 'My friend!' Job done, it was back to the jet to get airborne for Mombasa as quickly as possible.

My crew had been great. I have had crews where not only the passengers were whining, but the crew were too. To say thank you, I took them into Mombasa and we had a fabulous meal. We were there for a week, and when I got back and had to fill in a form to justify spending company funds, I wrote: 'The only way I can make this work is to have an exchange rate of one dollar to the pound.' No one said anything, so obviously they knew.

During my time as a first officer, captains all had a lot of experience even before they arrived at Britannia. The company examined each pilot on their ability to fly accurately, and to run the show and fulfil all the other responsibilities, but there was no training for the softer side of the operation. I had eight years in the right-hand seat watching captains dealing with things on their own. That would mean organising hotels, transport, whatever. In those days we were a European operation, and we were a well-known company, so we could sign for different services. If I had to pay out a big lump of cash, I could do that. You can train that aspect of the job as much as you like, but I've learned that some people can command an aeroplane and some struggle. Some can fly the jets and deal with the aeroplane, but when it comes to dealing with the crew, they struggle. When it comes to dealing with the passengers, they struggle even more.

One time, I had the whole crew in the back galley en route back from Abu Dhabi and told them: 'We've been away for three weeks, I've looked after you and you've all let me down. So here's what's going to happen: I expect to get a good service and nice coffees on the way back and a smile. And if I don't, I'll suspend you all.' The reason that got out of hand was a weak cabin manager who didn't control her crew and I eventually had to step in and sort it. I have even sent a crew member back to the UK. On the way out to Australia once, I said to one of them: 'You're going home.' She replied: 'Ha ha, no, I'm not.' I said: 'Yes, you are. You're not coming any further.' If I had a good cabin manager, they took a lot of that sort of problem away.

The importance of being earnest

From my earliest days at Glasgow Flying Club and then as an owner of my first light aircraft, I had always been taught the importance of safety and proper maintenance of an aeroplane, and of being sensitive to its performance, whether in the air or on the ground. That applies as much to a B767 as it did to my little wooden-framed Jodel. The insurance cover on my own aeroplane, the Cirrus, is for millions and I would only ever fly it with full insurance. Every year we have to declare how many hours we've done in the last 12 months, and someone less experienced coming into the group will have to pay an extra insurance premium.

The aircraft I learned to fly on all had to have a C of A (certificate of airworthiness), partly because they were being used for hire and reward when I was paying for flying lessons. The aircraft would be serviced regularly by qualified engineers to certified standards. Any work done would have to have a coherent paper trail to allow the Civil Aviation Authority to carry out an audit on all documents and an aircraft without a proper paper trail is almost worthless.

There is a formal check after every 50 hours of flight, which might only be a visual inspection and an oil and filter change, depending on the sophistication of the aircraft. The next check would be another 50-hour check, then after 50 hours it would be a 150-hour check and then back to a 50-hour check and so on until the annual check, which is irrespective of the hours flown. Small private light aircraft can be registered with the LAA (Light Aircraft Association) based at Turweston Aerodrome, where the aeroplane I currently co-own is based. The LAA issues a permit to fly, which means an owner could undertake some maintenance under guidance from the LAA engineers. With this backup, I did some work on my Jodel, although I didn't have her long enough to do much. There was a limited number of things I could safely do myself and I was very earnest in my endeavours. I knew there were areas of the aeroplane it was not safe for me to go near. This more specialist work was carried out by the chief engineer at Loganair, George Cormack, who was one of the group members. Nevertheless, with those little aeroplanes, in those days you couldn't leave the country or fly over towns – though it would be pretty hard to spot from the ground if a particular aeroplane had a permit to fly.

The Americans are a bit more pragmatic, and keeping an aeroplane on the American register is cheaper and simpler than the British one. The Americans gave me a private licence some years ago. I had applied for the single-engine PPL so I could hire a plane when I was in the States. Ruth had joined me on a trip and at the Federal Aviation Administration desk in Orlando, I was asked: 'Captain, what did you fly over?' I said a B767-300, to which he replied: 'I'll give you the multi-engine rating while you are here.' In the UK, I would need to do a course plus a test to achieve what this chap had given me for nothing. The Americans have a very sensible attitude to aviation. Over here, we make it as hard as possible - and as expensive as possible.

I had no idea how the DC-3 we encountered in America with Jock Adams would have been certified. All I knew was that Jock was getting in it and therefore so was I. Even if he had said 'Hold the Sellotape round there, Scott, while I lash this thing together', I'd have held it. In one of the DC-3s we flew from Entebbe, there were great big holes in the canvas elevator. You could put your hand through. One of them had ripped quite badly and had to be mended in Khartoum where we stopped. The old radial engines, like the ones fitted to the DC-3, were becoming more and more unreliable as the engine manufacturers increased the power by having multi-rows of cylinders, intercoolers and two-speed superchargers. The modern piston engines fitted to most light aircraft today are very reliable and complete power loss is very rare. An engine on a light aircraft has a TBO (time between overhauls) of about 2,000 hours depending on engine type and certification. After that, in some circumstances the engine life can be extended 'on condition' and up to about 10% of the 2,000 hours. If not, it is a total rebuild or buy a factory reconditioned or new engine. There won't be much change from £50,000 for a six-cylinder engine.

Piston-engined aeroplanes are much less reliable than a jet engine, which is simplicity itself. I have flown thousands of hours in jet-powered aircraft. I have even been in one which was hit by a bird. I was in the jump seat and even though the bird went into the engine, it didn't stop. We hit it on take-off, and the captain aborted, taxied in and shut it down. When the engineers looked at the engine, it could still run despite all the blades being twisted. There is a YouTube film clip showing a Thomson B757 taking off from Manchester and hitting a bird just as it is lifting off the runway. The engine stalls, you hear it banging and see flames coming out. The passengers must have thought they were all going to die, but the crew and Manchester ATC did a fabulous job. The chances of having an engine failure now are minuscule. It is estimated that during a career as a jet pilot, there are not enough engine failures occurring for us all to have one in all the years we fly.

If you look back to the days of the old Constellation and their huge radial engines, which were turbo compound 18-cylinder and supercharged, they were losing one a month. Jamie Alderman said when he was flying the Lancaster with the same engine as the Spitfire, a Rolls-Royce Merlin, it failed regularly. They had four of them to achieve the power required because engine failures were just normal. When they were flying the Connie, it was normal for Jamie to shut one down because technology was being pushed to its limits. These days, we never take off with full power if we can help it, but work out how much power we need and that's what we set.

With a modern jet engine, the definition of severe damage to an engine is a fire or a failure with vibration. N1 and N2 are the spools in the engine and N3 is the third spool in a Rolls-Royce Trent engine. The B767 GE CF6 is only a two-spool engine but if one of them was seized, that used to be deemed severe damage and the pilot would have to go through the severe damage QRH checklist. If the engine just wound down, you would do the QRH engine failure checklist, but not by memory. If it was severe, you would do the first few actions by memory, then call for the QRH to complete the rest by reference. That's not the case now. You can have an N1 seized and if there is no vibration, you would treat it as an engine failure. Jet engines are incredibly robust and reliable things.

There are films of an engine where a large bird is fired into its core at high power. It's worth watching. The engine starts to eat itself but nothing breaks free and comes off. What is absolutely required is that these engine blades do not come off and are not allowed to come through the fuselage into the cabin. The whole engine has to be contained within the engine case or exit via the tailpipe, and you see in the pictures the engine casing bulging out to absorb the forces. If you look down a modern high bypass jet engine, it appears to be quite hollow inside - 75% of the air is bypassed around the core of the engine. Each engine thrust on the B767-200 was about 48,000 pounds but the newer version, with uprated engines, had increased power to 54,000 pounds of thrust, with a fuel burn reduction of about 5%. More power for less fuel. On the B757, with Rolls-Royce engines, you don't climb at full power unless you need to, but on the B767, at 10,000 feet it always reverts to full climb. The B767, like all twin-engine jets, is so powerful because one engine has to have sufficient power to continue to accelerate down the runway, take off and climb away if the other engine fails at V1.

Most of the checklists are completed before the taxi out, only leaving the short 'before take-off' checklist . It often makes me wonder: 'How many times have we changed the way we fly this aeroplane?' The taxi phase hasn't really changed since the days of the Wright Brothers. Airports are a lot more complex and you can fly in much poorer weather conditions now than years ago, eg, fog. Everything has become hugely technological but the taxi out can still cause problems, as we saw when a BA Jumbo hit a building when taxiing down the wrong taxiway in South Africa. The aircraft was subsequently written off, an expensive mistake. Airport lighting has got better but ground radar is rare and it is still possible to taxi down the wrong taxiway, especially in airports such as New York, where there isn't even grass. It is just one huge concrete expanse, so it isn't uncommon for a pilot to get confused and taxi down the wrong way. Sometimes there will be a tiny note telling you that this taxiway is not suitable over a certain weight, or wingspan, which can be easily missed on the airport taxi chart.

In Birmingham, taxiway Delta, the inner taxiway, is fine in the B757 but it won't take a B767 due to its larger wing span. I've heard ATC say to some poor pilot: 'You're taxiing the wrong way, sir!' What Boeing is advising now is: 'This is a busy time. Maybe it would be better if you do flap selection and all these other things before you taxi so you can concentrate on taxiing.' It is now acceptable and encouraged after landing to taxi in on one engine. The aircraft is light, since the vast majority of the fuel has been burnt. You could be 60 tonnes lighter than when you took off, so you need a lot less power as you taxi in, but a B767 always wants to accelerate as it is so powerful, especially once you've landed. With both engines going, you will need to be continuously braking, and the brake temperature figures will be going up and up. If the brakes get really hot, there is a chance of a brake fire, then a wheel fire. There are temperature gauges in the flight deck and, once they get to number 5, you may have to think about telling Engineering, or certainly not trying to take off. Flying into Alicante or somewhere with a long runway, with the aircraft at normal landing weights, idle reverse thrust is fine. Gentle braking and using the available runway keeps the brakes cool. You can set the autobrakes at different deceleration rates through 1, 2, 3, 4 and max. Some pilots tend to taxi too quickly. They are impatient, so they put power on to get moving and then have to brake as they come to a turn. So they tend to be fighting the power and the brakes and when the aircraft is light, especially the B767, it just runs away. If you are riding the brakes to keep it at, say, 20 knots, they just get hot.

If the aircraft is heavy with freight, and the weather hot, as it is likely to be on landing in the Caribbean, you need to plan to use the whole runway with a low brake setting and use lots of reverse thrust to get on to the stand with the brakes still fairly cool. This is where single-engine taxiing is handy because once the engine has had its cool down - either one minute or three minutes, depending on the engine - you can then cut an engine, which will mean the brakes are not fighting the power of two engines. What you are meant to do is let the aeroplane accelerate up to 20 knots or, in a straight line, a maximum of 30 knots, then brake down to 5 knots, thus letting them cool between applications. Turnaround may be only 35 minutes, and if both engines are kept running during taxi in, the brake temperature could easily delay departure. In a B767-300, the maximum take-off weight is 186 tonnes and the lift-off speed is about 170 knots, which is a lot of momentum to stop – especially when you think of the energy of 186 tonnes going at 200mph. If necessary, the brakes are going to have to stop you by the end of the runway, which you are already two-thirds of the way down. Once you pass 85 knots, if you were to reject the take-off, at that point you would close the thrust levers, check the autobrakes are functioning, raise the speed brake and select full reverse thrust. The first officer would then tell ATC and call 60 knots as we decelerate. If the aircraft was light, as on a short-haul departure, none of this would actually happen because the aircraft would have stopped almost instantly and very violently.

There is no way of replicating that - a simulator just can't do it. When pilots do a 100 knot reject in the simulator, it will lean forward, trying to replicate deceleration, but only very approximately, which is not what it's going to be like in the aeroplane. If you are at a light take-off weight and reject at 85 knots, it's going to be like hitting a wall. Another procedure the simulator cannot replicate is a decompression at altitude. When crew go into a decompression chamber, they can pass out due to hypoxia. We leave that to the military. In the simulator, there will be an aural warning and warning messages. There will also be a loud roar, replicating outside air flow but no physical effect at all, there is no misting in the flight deck. In a real plane, you will have air trying to get out of every orifice in your body so neither the pilot nor the crew must ever fly with a cold, as during a decompression, the pain in the sinus or the gut would be severe.

Introspection

It was with great sadness that I learned of the death of my old mentor and inspiration, Jim Beaton, in a flying accident in the United States at the age of 78. Jim had been engaged by American King Air Services, in South Carolina, to take a new single-engined Cessna 172 Skyhawk from the factory in Kansas to St Petersburg in Russia, a distance of some 5,000 miles. He completed the first leg from Kansas to the international airport at Bangor Maine on the east coast of America. Then bad weather set in and he was forced to sit it out for three days, but on the fourth day the weather report showed improvement. The Cessna had been fitted with one extra-large fuel tank capable of holding 125 gallons for the Atlantic crossing which, theoretically, would increase the Cessna's range from five or six hours' flight time to around 17 or 18 hours. He set off bound for Goose Bay in Canada, the next leg of the crossing, on to Reykjavik in Iceland, Prestwick in Scotland, and then to St Petersburg, where the aeroplane was due to be received by the city's university flying club.

Within half an hour of Jim's departure from Bangor, he was on the radio to ATC reporting severe icing on his wings as he tried to climb through the icing level. Eventually, he told Bangor he was turning round. His initial plan was to land back at Bangor. Soon afterwards, he reported that the aircraft was in severe trouble and losing height. He was going to make an emergency landing at Old Town DeWitt Field, a small municipal airfield. The wreckage was found some 10 miles short of that airfield. The Cessna's emergency transponder deployed, and search and rescue found the plane many hours later, half submerged in a frozen river. They took Jim's body from the cockpit and the view was that he had died on impact. Questions immediately arose about the factors which contributed to the tragedy. Was the single-engined aircraft carrying too much weight for the flying conditions, with its extra fuel capacity? Should Jim have attempted the flight in the first place? Was he misinformed about the severity of weather conditions? Had he been attempting a forced landing on what might have appeared to be a snow-covered frozen river? Opinion afterwards suggested that the aircraft had been at full power as Jim tried to keep it in the air, but the ice build-up on the wings was destroying the lift and it was descending even with full power on the engine. There was also a report from one of the search and rescue directors, who knew Jim, that while waiting in Bangor Jim had been anxious about a sick family member and was keen to get home.

Jim qualified as a pilot in 1955 with the Scottish Flying Club at the old Renfrew Airport I used to fly from in my early days. In 1968, he qualified as a flying instructor and helped to set up Glasgow Flying Club. Later he went on to teach aerobatics, seaplane flying, and 'tail-draggers' – planes with a castoring tail wheel - which were to become his great love. Jim moved to Gloucester in 1980, where he worked as a pilot for 10 years, flying a jet for an international businessman. He eventually went back to his first love, flying light aircraft. He held European and American licences and, even at the age of 78, was still passing his six-monthly medical examinations. Hearing of an incident like this, where an incredibly experienced and knowledgeable pilot loses his life, possibly as a result of misinformation or poor judgment, is bound to pull you up sharply and get you questioning your own approach to flying. Are you cutting corners out of familiarity or complacency?

The news of Jim's death was a real shock because there is no doubt that I wouldn't be a pilot today if it had not been for him. There was added sadness because that old killer which pilots know as 'get-you-home-itis' might, and I stress might, have been involved in his death. Jim had beaten it into my head in the early years that the desire or need to get home, or on to the next destination, had killed more pilots than anything else in aviation. If you asked a pilot in the bar 'Would you fly in these conditions?', they would confidently say no. But next morning, with a meeting to get to or a problem at home to sort out, they might well leave their brains and judgment on the ground and launch into the very weather conditions they had said they would never fly in. We are sometimes reminded that Mother Nature is bigger, scarier and more powerful than we are, and the ability to say 'No, I will try tomorrow' is the decision many a dead pilot should have made.

What I did throughout my career in Britannia was the same as I did from the beginning, as a young private pilot. Once I had got on top of flying the jet and settled into the right-hand seat, I wanted to move up from line first officer to training first officer, and then command. I couldn't believe I was being successful at this fantastic job. My career was full of ambitions to chase after, to achieve and to enjoy. Most of them fortunately involved more money, so this was always ticking the right boxes. I was elected to the British Airline Pilots' Association (BALPA) company council by my fellow pilots and I felt very proud of that. I think that single-minded focus might have been part of my continuing search for self-worth and fulfilment. The vast majority of pilots don't strive for all these roles during their careers. They perhaps become a captain and think: 'That was a big hurdle but I've made it.' Then they get on with life.

Flying is an adrenaline thing as well. I enjoyed flying a B767. It's a buzz, there's no two ways about it. I looked forward to the occasional long-haul flight, but not to being away from home for long periods. Wherever I am going, the flight is still exciting - that doesn't go away. With long-haul flights, I was coming back feeling jet-lagged for a day or two and I would think: 'I've had enough of this.' In the early days, large parts of the job made me smile, including flying with captains who were fun to be with. I guess I became the old captain, and I tried to run a relaxed and friendly flight deck.

When you are sitting in an aeroplane for nine hours, you always talk about life outside the flight deck. Cabin crew girls and boys come in and out, and the chatter is the usual: 'What are you doing tonight?' By the time you arrive and go on to local time, it is usually late afternoon. You can have a drink, but you don't want to go on a bender. Historically, if I drank too much on the first night, I would get a terrible headache even after just two beers. I used to fly to LA in the Jumbo and it is eight hours behind the UK. The first time I flew there, the flight engineer took us for sundowners on the top floor of a beautiful hotel overlooking the Pacific. We drank quite a few margaritas because we had a clear day the following day. It might be 11pm where we were but it was now 7am as far as my body was concerned - and I'd already done a long flight. I went to bed and woke up with drilling in my head! It took the whole day to return to any semblance of normality. The crew were going out for the day and I said: 'I can't.'

The next time I flew to LA, I told myself I wouldn't do that again, so I just had a couple of beers – and woke up the next morning with someone drilling in my head and the whole performance over again.

The next time I went to LA, I took no chances at all. I just went to bed – and I woke up with someone drilling in my head. I thought: 'What the f***...?' I saw a pharmacist pal and told him that every time I went to LA, I felt like shit. I said I'd get him a first-class ticket, we'd share a room, and he could tell me what was going on. He came with me and just watched me the whole time. He said: 'You're not drinking enough water. That is the core cause.' What most people don't know, and what pilots and crew often forget, is that the air inside an aircraft is dry – often drier than most deserts. It is much cleaner even than an operating theatre, but it is unbelievably dry. I have been told that you can't drink quickly enough to replace fluid in the body. After that, I always carried a two-litre bottle of water with me in the flight deck. You have to force it down - if I didn't finish that bottle on any flight, I knew I wasn't making an effort. The B767 flight deck has its own air supply. The cabin air supply is normally mixed with recycled filtered air. All the air is drawn out the back of the aircraft through the outflow valve. The quality of air is not an issue on the aeroplane. When people bang on about it being recycled, some of it is, but not all of it and the filtration is very effective.

The approach into LA, after a long flight, is demanding because you are descending over the mountains, so you are kept high, and the RT is very busy. Americans talk quickly with a slightly different RT style. Most American pilots only fly around the US and they become used to local ways of RT speech. Their language is clipped and if you ask them to repeat, they get impatient. In the US, if the airspace is busy, ATC tend not to send you into a holding pattern as they do in the UK. They start to radar vector the aircraft to get them sorted for the approach a long way out. It is completely normal to arrive into Gatwick first thing in the morning and be told to take up the hold at the Mayfield beacon for an approach in 10 minutes. When I flew into Orlando, I don't think I was ever held, but ATC started radar headings maybe 200 miles out: 'Take up a heading of so and so, reduce to 220 knots.' Unlike in the UK, their radar screens show the large storms and they'll take you round the weather and feed you in for the approach. Air traffic radar in the UK is designed to remove the storms from their screens, so onboard radar screens show up the storm cells and it's pilots who say: 'We need a left turn 30 degrees.'

My pal the pharmacist sorted me out – 'drink more water, take it easy on the booze on the first night, and as soon as you get any form of headache, take something.' I hardly suffered again. When it happens, I get to bed asap and try and sleep with someone drilling in my head. It is way past a heavy head so whatever it is, I am just wiped out. Someone can say 'We're on fire!', and I'll say: 'I'll just burn.' My brother, who flies for BA, makes a point of staying on UK time when going to the United States. I went to a conference in LA once, where the head of the CAA was the keynote speaker. He had just flown in for that speech and then flew straight back. He said he met a few British people wandering the hotel corridors at night because they were staying on UK time, and I thought: 'That's a novel thing to do.' But it's a hell of a lonely, miserable life! There are two negatives to staying in UK time – it is completely anti-social, and it can make the return flight an ordeal because you will be tired. If you come in from your flight, have a shower, and go straight to bed, that might be fine - but then you wake up at 3am. If you get up at 3am, your next flight back to the UK is 9pm, so not the best time to go to work if you have been up all day. By going into local time, I could wake up at 7am, which is lunchtime at home. That meant I had a good old lie-in to help me through the next night for the flight back.

On one Atlantic crossing, we were really struggling. We were a crew of three coming back from Cancun. Ben Bamber, my pal who was a Civil Aviation Authority inspector, was with me flying the jet, but I was commander. I was sitting in the left-hand seat and the first officer said: 'I've dropped off. I've just had five minutes sleep. I'm going to put in a fatigue report.' The point was that all three of us were knackered. We had done nothing different to any other night stop. I had done a similar flight the week before and it was fine, but sometimes your body reacts differently. Some of those night flights, back over the ocean, can be incredibly tiring and so can dragging back at 3am from Paphos. Usually the adrenaline starts to kick in from top of descent and keeps you going through the landing. Hopefully, that also gets you through the drive home.

To have or to hold

Clearly, the sleep issue for pilots is a very serious matter. So serious that I was sent off to Paris with one of the cabin crew managers to attend a course on sleep and fatigue with NASA experts. They were world-renowned experts on sleep, especially, of course, for people in space. I spent a whole weekend there and it was very interesting. Our chief pilot at the time had said: 'If you are on a long haul and you can't sleep, just lie in the bed and shut your eyes. That will be as good as a sleep.' I mentioned this to the NASA expert and he said: 'Gee, son, that's like showing a starving man a Mars bar!' I had great pleasure going back to our chief pilot to tell him that. We covered things like drug and alcohol use, sleeping patterns, the difference between being in a light or dark room, and circadian rhythms. We also dealt with the benefits of cat-napping for 20 or 30 minutes. I wrote a half-hour module to be included in the winter joint crew training programme – but the company refused to let me include it in the course. I guess it was too much of a hot topic.

Various countries have come up with rules for levels of alcohol in the blood. In the UK, it is 20mg per 100ml of blood. A captain was reported to the police after he checked in his bag before a long haul-flight. Police knocked on the crew room door, he was arrested and breathalysed, and the company immediately suspended him. While checking in his bag, the person behind the desk had smelt alcohol on his breath. After two lots of blood tests and a month off the line, he was found to be below the limit for flying and returned to the line. We had another incident in Orlando. The security people smelled alcohol on the captain's breath and he was taken off the jet by police and arrested. The whole operation was grounded and the newspapers were soon on to the story. It turned out he had been on a special diet and that had caused the alcohol smell on his breath. I know there have been incidents of pilots and cabin crew drinking on duty. This kind of alcohol problem is usually hidden, and only found out after someone is reported or caught in the act. The culture has changed since the days of Jamie Alderman.

In the medical section of one of our manuals it used to say that alcohol affects you, which hardly requires a rocket scientist to figure out. There was just a wishy-washy chapter, not telling us much. These days there is more information on acceptable levels of alcohol in the blood but no really useful advice. I still hear pilots saying: 'A few beers will help me sleep.' The NASA experts say: 'Yes, but that will be rubbish sleep.' They went into great detail about what alcohol does to your sleep. Most people knew that when we got drunk at a party, next morning we would wake up and be knackered. It's easy to just think: 'I dropped off as soon as my head hit the pillow.'

Another revelation for me was that a catnap was very powerful. That is something that has since been established and pilots are now allowed to have a catnap on the aeroplane. NASA stated categorically that a short nap of 25 minutes will revitalise the pilot, but if you sleep much longer, it can result in sleep inertia, where you wake up feeling groggy and disoriented. Therefore, it is now acceptable that if one pilot wants to shut his eyes, as long as the other one is wide awake, they can have 20 minutes. I found it incredibly difficult to do. I even tried earplugs and putting my seat back but I could still hear what was going on. Some people drop off like a baby. If I was doing a quick turnaround, I tried to have a good sleep during the night I was away and get up as late as possible on the day of my return flight. I had little alcohol and lots of fluids. Sometimes I tried to have a doze in the afternoon but it usually didn't work. I had a few coffees and then went on to lemon tea. Too much caffeine can be quite negative.

The training department wasn't directly involved in sleep issues or fatigue reports. The fleet office and pilot office dealt with the day-to-day running of the operation, and pilot welfare and behaviour. They decided how we flew the jet, and the training department's job was to train the pilots.

We know about duty hours because that's a legal requirement. We know you can phone up and say you are fatigued and the company will have to accept that. And we know there is a difference between tiredness and fatigue. It is only in recent years that fatigue reports have been much more widely used. BALPA has encouraged pilots to fill in a fatigue report every time they feel fatigued. Unless fatigue reports are submitted, there is little the CAA can do when they are in contact with the company. If they are getting lots of fatigue reports over a certain duty or a roster line of duties, the company will need to take heed. Now people are putting reports in, as my young first officer did on that flight over the Atlantic. This is a good thing because at one point people thought: 'If I'm putting in a fatigue report, I'm popping my head above the parapet.' With BALPA encouraging them, they feel they are not on their own. The company missed a great opportunity to explain to pilots and cabin crew the simple fact that, at four in the morning and four in the afternoon, the human body wants to sleep.

On the B787 there was a bullet – a return flight with just a short rest - to Phuket from the UK. When we went to Phuket in the past, we either night-stopped in Abu Dhabi, if the flight originated in the UK, or flew direct from Scandinavia, but always with a few days off in Phuket. The reason for doing something like the Phuket bullet is that someone in the scheduling department thinks 'We can try to get the crew back', because they hate the idea of well-paid pilots lying on the beach. In most airlines, the aircrew are not liked because they are the ones away having fun. They are licensed and work to legal rules, and are overpaid compared to many in the offices. Management would deny this with their dying breath, but I have listened on many occasions to managers banging on about overpaid, over-opinionated pilots who are hard to manage and keep in line. Sadly for them, they need pilots.

Another safety issue came up when I was positioning somewhere in the Caribbean with my crew on a private jet and one of the cabin crew was wearing flipflops in the aeroplane. I said: 'Wear what you like, of course there are no rules, but imagine this aircraft coming off the runway on take-off and there are bits of hot metal and debris all over the place. You are going to run out of the aeroplane, slide down the slide, and with a bit of luck make your way through a debris field in flipflops.' She said: 'I had never thought of that.'

More and more, with modern air travel, the whole concept is pretty much to turn the aeroplane into a bus. In military transport aircraft, all the passenger seats face backwards because it was established in the Fifties you were ten times more likely to survive a crash in a backward-facing seat. On civilian aeroplanes, we don't do that and all the passengers face forward. If you turned the passenger seats around, you would still get an impact, but your head and neck wouldn't be thrown forwards - they would be thrown back against the seat. Not many aircraft reverse into mountains. In some of the old aeroplanes, there was a mixture of front and rear-facing seats. Airlines don't do that now because if they did, the passengers would all ask: 'Why are you doing that?' Of course, the answer would have to be: 'It's safer.' That would simply prompt the question: 'You mean the aeroplane's not safe?' They don't want to imply that there is any lack of safety. I won't even introduce the subject of the benefits of a three-point seatbelt. So the whole flight experience is being turned into something akin to a bus ride and you end up with your own crew member wearing flipflops. There is another safety issue with hand luggage.

I do not understand why it is made financially advantageous to stuff a bag – as stuffed as you can get it – then bring it into the cabin and cram it into the hat rack. These sections of the cabin can fail in a crash and those cases turn into missiles whizzing round the aeroplane. The overhead lockers are designed for hats and coats. The hold is where larger, heavier things should be. These days, you usually have to pay extra to use the hold, which is criminal. The CAA should simply say no. I find it appalling that no one has the balls to do so. What happens is that the whole aeroplane becomes overweight because these bags are not usually weighed. They are given a nominal weight, which is okay when the passenger has all their heavy stuff in a case in the hold. Now they have to pay to put a bag in the hold, passengers cram as much stuff as possible into their carry-on bag. Surprise, surprise! In any kind of incident there is likely to be severe deceleration, and when that happens, people will wish they didn't have their heavy bag up there above their heads.

Airlines are trying to make extra money at the expense of flight safety. Yes, they are fulfilling all the regulations, but they are too money-led to say: 'What we are doing is madness. We should have a limit – ladies' handbags and nothing bigger.' It is called a hat rack for good reason.

Keeping in touch

When I started flying to Australia, I used fax to keep in touch with Ruth and the family. I found that much better than calling by phone. I'd arrive in a hotel in Singapore and there would be a fax from the girls or Ruth waiting for me. Then we went on to text.

In 2003 I flew the Hajj, where Muslims go to Mecca once a year, and we were flying them from a few places in Indonesia to Jeddah. It was Britannia's winter season, which was always a quiet time, and any new work coming our way would help the company cash flow. Garuda Airlines awarded Britannia the contract to fly the operation with our B767s. Five or six B767s were based in either Jakarta or Ujung Pandang on the island of Sulawesi and they would fly via Singapore or Batam, south of the straits of Singapore, and then on to Jeddah. The aircraft would run back and forth, for about a month, flying out full and returning empty.

The religious festival lasted about ten days and it would take another four or five weeks to get everyone home to Ujung Pandang. We started flying the Hajj in the mid-1990s but in 2003, because of the bombs in Indonesia, Britannia decided it was not safe to base our crews there. They moved to Singapore and I thought: 'I'll have a go at that.' I did that Hajj and we were based there for about three-and-a-half months. They put us up in the Grand Hyatt, where we all had junior suites and I kept my suite for the whole time. If I left to go away for a few days down route, I would come back to find all my washing laid out on my bed in lovely wicker baskets and tissue paper between all my shirts. Of course, when I came home I couldn't help pointing out to Ruth that she didn't do that for me. The black eye eventually healed. Ruth and Alex, my fourth and youngest daughter, came out for the break in the middle when the festival took place. The rosters were written to give us all two weeks off. Some pilots went back to the UK, but most went off on trips, making the most of being the other side of the world. Alex came out with Ruth because it was the Easter holidays, and we had some time together in Singapore. I was flying for the first couple of days of their stay and Ruth and Alex spent time with the other families who were out there, then we all went off to Perth in Australia and spent a week driving south along the coast in hired motorhomes. I had done three Australian trips in consecutive winters, so I was familiar with this part of the world.

I was using fax to keep in touch then, but by the time I flew the Hajj, the company had issued captains with mobile phones so we could use text. I had no idea how. My first officer had to show me. Texting was far more convenient than fax and it became the favoured method of communication for Ruth and me. Now, of course, we have email, Skype and FaceTime. When I did a round-the-world trip, I Skyped home once or twice but it wasn't really our thing. More recently, as soon as I landed somewhere, I'd text to say: 'I'm here.'

By mid-1984, I had 3,500 flying hours to my name, of which 1,500 would have been on the jet. I was still a first officer and the emphasis was on me learning the skills to fly the jet. I was starting to crack the flying side, so now it wasn't an anxious time every time I got into the aeroplane. I could cope with everything in my role as a co-pilot, flying the B737 around Europe and going away for three to six days at a time in the UK, usually with a crew. That was my life, flying with all kinds of people, and I was getting to know the routes and the airports.

The airport at Naples, for example, has always been a difficult approach because, first, you are dealing with Italians, and, secondly, it has an unusual glide path angle. On top of that, Mount Vesuvius is very close, as well as high terrain on the western side of the airport. In addition, the whole approach procedure is complicated and difficult to fly, and there are lots of different approaches. There must be someone in an office in Naples who makes money designing all these approaches because the airport shouldn't be that complicated. There are about twenty different arrivals and approaches, and of course you don't know which approach and arrival you are getting until you've started your descent. Sod's law would dictate that it won't be the one you have programmed into the FMC (flight management computer) and briefed the other pilot on before TOD (top of descent), so the rush to catch up begins. There is always a sense that you are scrambling around. Coming in on the easterly is much more demanding because the runway slopes and there is high terrain close in on the approach. That means when you are on final approach, if the runway slopes up, you always look high, and if it slopes down, you look low. Before that, you will have come over some terrain. It will have been a non-precision approach, which is not as accurate and has a higher minima (decision altitude) than a precision approach like an ILS (instrument landing system). All of this added up to being very tricky – especially when the Italian factor was added into the mix. They have a one-way system on the taxiway, which isn't at all obvious, and sometimes ATC direct you to taxi the wrong way around. They just seem to make it difficult.

I was there in later years in a B757. The captain was on a final line check because he was returning after a few years away from flying. He had completed all the ground and simulator training and his line training. I was in the right-hand seat as training captain because he couldn't fly with a co-pilot until he was line-checked. As part of the line check, the captain had to operate as pilot flying and pilot monitoring, so I had flown into Naples and he was pilot flying on the way back. Between starting engines and the runway holding point, the captain had demonstrated that he was not ready to fly the line. Naples had beaten him. He didn't realise at that point, but he had failed. I was pleased that the shambles of Naples had caught him out because, by failing, it gave him the opportunity to fly a few more sectors with a training captain to sort out the issues that led to a fail. He was flying the line just ten days later, after being rechecked. Places like Naples will bite you if you give them a chance. It is not the sort of place just to arrive and try to sort out. It is best to take the time on a previous flight and have a good look at the charts.

On the B737-200, the instrumentation and navigation were basic. The 200 first came out in the late 1960s and by the time the 1980s came around, Britannia was still getting brand-new B737-200s but paying to get the latest navigation equipment taken out. All the good stuff was coming out of the aeroplane in the 1980s to make it the same as the very first 200 we had in the 1960s! The idea was to make all the flight decks the same, so each aircraft was identical. It was a great idea but technology was making those aeroplanes obsolete. When I went to Virgin, you'd get in different aeroplanes and they were completely different. Some had round dials for the engine instruments, some had vertical dials, and different aircraft would have different engine types on the wings. Your hand couldn't automatically locate a particular knob, button or switch because on one jet it would over there, but on another jet it would be over here. That is introducing 'cock up' because under pressure you can move the wrong switch. In Virgin, we had the luxury of the flight engineer setting the flight deck up and initialising the navigation system for us while we were having a cup of tea and a sandwich. That made things easier because they knew each aircraft inside out. There were six old Jumbos when I joined Virgin but they were all completely different. Britannia was trying to do the right thing, but we were flying these jets in a sense with one arm tied behind our back. If the captain said 'What's the ground speed?', I'd have to get a stopwatch out, time over six seconds and then say: '410.' Even the Trislander was better equipped than that. Having said that, Britannia was known for being top rate, punchy and slick. We were whizzing round Europe, working hard, the standard was good, the training was good – and, of course, the hooliganising was good.

Around 1984/85, Mum became quite seriously ill. At the same time, the company was looking for volunteers to move to a few of the out stations for the summer season. Those volunteering would be paid extra but the company would end up saving on hotels and allowances. Glasgow was one of the bases and I applied, but the request was oversubscribed so they used seniority to decide who would go. I was too junior to get the posting. However, once I told BALPA and the company that Mum wasn't well, they agreed to add me to the list.

Alan Roebuck was a captain, an ex-Phantom pilot who was Glasgow-based. I got quite pally with him and he was a lot of fun. As it was the beginning of the summer season, there was a new girl on the aeroplane, and in those days it was the habit for experienced crew members to give new girls a hard time. They might tell her: 'You need to take this bag of duty-free down to the rear hold because the husband wants to use his late wife's duty-free allowance.' Off she would go to explain this to the baggage handlers, who were in the process of loading the dead woman's coffin.

In the B737, at the back of the central console, there was a knob with 'Brake' on it. It had nothing to do with the aeroplane's brakes, it was to do with the trims – although it was never touched during the flight. On one of my flights with Alan, we had a new girl on board and we said to the Number One: 'Announce that there's an emergency and send her to the flight deck. Say we need her and tell her that the senior crew need to stay out the back'. Alan had fiddled with a few switches, maybe turning a pump off, and as a result there were a couple of lights on and corresponding 'Master Caution' lights showing. When the new girl arrived, he sat her down on the centre seat and told her sombrely:

'We've lost this and we've lost the brakes and you need to strap into the central seat.' This seat folded down because the B737 has a tiny wee flight deck. 'When I shout brake on landing,' Alan continued, trying not to laugh, 'you need to pull that knob as hard as you can because you will be responsible for braking the aeroplane.' We arrived at Glasgow and Alan landed the aeroplane, with a bit of a bump as it happened, and as he did so he shouted: 'Brake!' The poor girl began pulling on this thing with all her might. When we came to our normal taxiing speed, we turned around in our seats and she looked as if she had been attacked by wolves. It was only when we all ended up in the bar later that she was told about the prank. She was a good sport and took it well. I guess she didn't have a choice.

There is a story of a captain emerging from the flight deck, walking backwards and feeding out a length of string in each hand. He slowly moved backwards into the cabin, occasionally pausing to control the aircraft. As he pulled one of the strings, the aircraft rolled in that direction. After levelling the wings, he then continued moving slowly backwards. On reaching the back row, he stopped again and pulled one of the strings. Sure enough, the aircraft banked slightly. Looking down at the passenger sitting in the aisle seat, he hesitated for a moment before handing him the two strings and telling him, with a deadpan face: 'Keep her flying straight while I dash to the rear loo.' Before leaving the flight deck, the captain had tied the two strings to the first officer's epaulettes and said that, when he pulled on one side, he should bank the aircraft in that direction.

When I was first officer and flying the B737-200, I always enjoyed flying with that New Zealander captain. Sitting on the ground at Luton before engine start, he suggested we play a little game on the way to Corfu. He said whoever could say the word 'Barcelona' the most times on the PA to the passengers would win. The flight to Corfu goes nowhere near Barcelona, so the game was on. He was pilot flying so it was his PA on the outbound leg. With the passengers on board, he picked up the hand mike and, with a big grin, said: 'Ladies and gentlemen, welcome on board. We are fully ready to go but we are number two to the aircraft beside us, which is going to Barcelona, and when the aircraft going to Barcelona pushes back, we will be off. However, if the Barcelona aircraft is delayed…' We hadn't even moved and he had managed to say 'Barcelona' at least three times. As soon as we started to taxi, he picked up the mike and continued: 'Ladies and gentlemen, we are following the Barcelona aircraft to the runway holding point. And once the Barcelona aircraft departs, we are the next aircraft for take-off, as long as the Barcelona flight isn't delayed.' Pause... 'Yes, the Barcelona flight is lining up and when the Barcelona flight....' At that point, the flight deck door opened and the cabin manager demanded: 'What the f*** is all this about Barcelona?' By now, I was laughing uncontrollably, tears were running down my face. I didn't even try to beat him on the way back. He was a star.

The job was becoming distinctly easier. I had sorted out my descents and landings, and the only thing I really needed to concern myself with was who I was going away with on these trips.

The pilot-less Jumbo

From my very first flight with Jim Beaton, it was made clear to me that the handover of control in an aeroplane is something to be done carefully and methodically. When Jim wanted me to fly the aircraft, he would say clearly: 'You have control.' I would reply 'I have control' - making it clear who was the pilot flying. It is exactly the same procedure when flying a jet. If the captain, who is the pilot flying, wants to speak to the passengers or to the cabin manager, he hands over control to the first officer before going on the PA or inter-phone. He might also want to study an approach chart or look up landing distance required for a particular configuration. Many times, I have had to sit and watch in the simulator while no one was flying the aircraft because they both thought the other pilot was the pilot flying.

Pilots are under stress in the simulator, running drills and checklists they don't do on the line, under check, trying to communicate with different people and agencies. Managing all this activity takes skill and requires everyone to be vigilant. If the captain was the pilot flying at the onset of the problem, once the aircraft is under control, he can remain pilot flying and let the pilot monitoring communicate and research the approach. Or he can hand over control and take control back as often as he wants, or he can communicate and read whatever he wants while the other pilot flies. The three methods are all acceptable, though the second option is the one used by most pilots, partly because they think since they are under check they should be flying the jet. But it is a two-crew aeroplane and what I tried to show them are the skills they can take out of the simulator and use on the line.

I liked to hand over control as soon as I could, so I had time to make a plan, share it, look up the manuals, and speak to people, including the passengers. When everything was done and we were ready to make our approach, I would say: 'I have control.'

The downside to any verbal communication is that under stress that communication tends to break down. I once experienced a near-miss with another aircraft over London. At the point where the jet was not doing what I wanted it to do, and the first officer was not making it do what I wanted, I called: 'I have control.' She didn't hear me. The other captain on the jump seat heard me clearly, no doubt because he wasn't so stressed by the situation. I took control by disconnecting the autopilot and hand-flew the jet. I banked quite hard to the right and dropped the nose of the aircraft. Initially, I could feel the first officer fighting me on the controls until I repeated my command and she realised I was flying the jet. I wasn't at all cross with her because my message just hadn't arrived in her brain. I know that the level of arousal in a situation like that can go from low and comfortable to over-aroused with an inability to listen, speak or function, all within a second. We just have to deal with that.

When I first took the controls of a jet aircraft after I had finished the simulator training, we had to do base training because simulators were fairly new and not realistic enough to be acceptable to the CAA. Of all the training roles, this is the most demanding and dangerous. We had to fly two-engine circuits. We had engine failures, with the training captain closing the thrust lever after V1, and single-engine approach and landings. We also did night visual circuits. There was acceptable increase in risk, because pilots had to train on the jet before starting line training with passengers.

When I tried my first V1 cut in the simulator, I crashed spectacularly, but at least it was only my pride that was hurt. Jets are much harder to control after a V1 cut than the old propeller aircraft, mostly due to the swept wing and very powerful remaining engine. Simulators had to be developed to enhance pilot training in this area, but in the 1980s and 1990s, base training was the norm. The consequence of this was that there were occasional losses. An Air Anglia Fokker F27 turboprop crashed while practising asymmetric circuits in the UK. When we first got the B767, we were still doing asymmetric base training and we nearly lost one of our new B767s.

It happened at Shannon when, just after take-off, the trainer brought back the power on one engine and the student put in the wrong rudder. The aircraft banked 90 degrees and the wing tip actually skimmed just above the grass until the training captain corrected the rudder and brought up the power on the throttled-back engine. By then, air traffic had witnessed the drama and hit the crash button. The crew got a real fright. The rules were changed by the Authority and now if you have to go base training, there is no single-engine work at all. It is covered in the simulator. On smaller passenger aircraft with no simulator available, asymmetric base training is still carried out.

It's not only flight control dramas like that one that pilots are learning to improve and predict. The whole operation of carrying hundreds of strangers from one part of the world to another will always provoke unexpected events which captains are continually learning how to deal with. On a night flight, eastbound over the Atlantic to the UK, the Alert call went off and the crew were 'on stations'. The cabin manager called to say a passenger had collapsed and a doctor was with him. The flight continued but the pilots began looking at different options in case an in-flight diversion was required. The doctor declared that the passenger had to get to hospital as soon as possible, so the decision was taken to divert into Lajes in the Azores. During the descent, the messages from the cabin were not encouraging and when they received the latest landing conditions for the Azores, they weren't encouraging either. Heavy rain and strong crosswinds were reported. By the time the medical team got on board, the passenger had died.

At that point, the important question was: are there enough flight duty hours still to make it to the UK? If they were able to refuel and turn around the aircraft quickly, then cruise at a high speed, they would just get back to the UK before the pilots ran out of hours. The airport commandant at Lajes came on board. It was becoming clear that the dead passenger had been a drug mule, and one of the packages of drugs he had swallowed had burst inside his body and killed him. The crew had identified his companion, since drug mules always travel with another person who is supposed to look after them. They began to get suspicious when the passenger did not eat or drink anything during the flight. Drug mules do this as, if they have to go to the toilet, they would pass what they were carrying inside themselves and would then have to swallow it again.

There was a discussion about the amount of paperwork that was about to be generated because of a death in-flight, and the consensus was that if he just happened to die in the ambulance after leaving the aircraft, all that paperwork would conveniently disappear. So, refuelled and with a new high-speed flight plan, they launched for the UK. On arrival, the aircraft was met by a large team of police and drugs unit personnel. Everyone was kept on board, and the passenger travelling with the deceased was arrested and the aircraft thoroughly searched. Just another example of how a nice quiet flight home can suddenly turn into a long, stressful flight home.

Of course, the personal lives of pilots can also introduce risk to the successful outcome of any flight. I've mentioned the vexed subject of what pilots call 'get-me-home-itis'. The syndrome continuously rears its head in aviation. As airline pilots, we can fly in most weather conditions but there comes a point where, no matter how much confidence you have in your aircraft, you will need to call it a day. I know from experience that I tended to be cautious. When I shot an approach in bad weather conditions, I was very comfortable with the thought that I might not get in and would happily head off somewhere else. If I was returning to the UK and the forecast was foggy, for example, and we were planning on a CAT 111 (3) approach – an auto land with no altitude minima and only 75 metres runway visual range (RVR) – I would carry as much fuel as necessary to divert somewhere where the weather does not require an auto land. I wanted to avoid an aircraft failure, which precludes auto land. This is not rocket science. I sometimes thought that if I went out into the cabin and asked the passengers to vote on plan A or my Plan B, they would all vote for the one that was most likely to ensure that they would get down on the ground alive.

A UK-based aircraft had crossed the Atlantic and was making an early-morning approach into Bristol, where there is a short runway built on a hill. The weather was wet, with low cloud and strong crosswinds. The captain, who was pilot flying, made the approach and hit the runway so hard that the landing twisted the fuselage. At one point, the aircraft was going to be written off but eventually, after consultation with the aircraft manufacturer, they erected a temporary hangar over the jet and fixed it. It was put down to 'one of those things'. Mother Nature can do all kinds of weird things to an aeroplane, and even the soundest of planes can be ripped apart by a sudden strong gust of wind. At some point, the pilot has to use good judgment about whether it is a good idea to try to land. It is never clear cut. If it was, it would always be easy. I sat at the controls very comfortable with my yellow streak on my back, knowing that perhaps it would one day save the lives of me and my passengers.

Vigilance is always required to prevent pilots being remote from their jet, and to get them thinking about their reliance on following the magenta line to a blue circle, which tends to get them thinking: 'What a good job I've done.' In my many roles, I have been lucky and privileged to meet some very clever people. One chap, Jack, a test pilot with Boeing, used to come over to visit us in Luton, and on one occasion we were talking about pilots becoming remote from the jet. He told me that Korean Airways had requested him to go over and fly with them for a time. He hadn't really fancied the idea and put them off several times. Finally he relented, and went over to fly their B747-400. He was the extra crew member so once the other two pilots had chosen their rest periods, Jack was left with first rest. He didn't feel at all tired so went to the crew bunk to read his book.

He awoke to find that he had overslept into the first officer's rest period by 15 minutes. Feeling guilty, he hurriedly dressed, had a quick wash and hurried out of the bunk area. As he passed through first class, he spotted the captain having his dinner. He went over and apologised for being late, but the captain brushed it aside, saying: 'Don't worry about it. It's no bother at all.' He carried on up the stairs to the top deck and, as he approached the flight deck, the first officer came out of the toilet. In total amazement, Jack asked: 'Who is flying the jet?' The first officer replied calmly: 'You are.' Needless to say, on landing, he caught the first flight back to Seattle.

The point of this sorry tale is that the first officer was happy to leave 400 tonnes of metal hurtling through the sky at 37,000 feet and 550 knots. If an engine had failed, the autopilot would have done a brilliant job trying to keep the wings level, but would have increased power on the remaining engines until it was so out of trim that it wouldn't have been able to hold it. At that point, the autopilot would have disconnected and the aircraft would have rolled on its back.

We sit in our warm, comfortable aluminium tubes whizzing through the air, but just one inch away outside, the temperature is minus 65C, there is no breathable air, and we are seven miles above the earth. There is no greater reason for telling every pilot in the sky: 'Get a grip!'

Britannia's newest captain

During this busy period, in 1985, I felt it would be a good idea to get back to the controls of a light aircraft, which, after all, had been my first passion in aviation. To do this, I flew with a chap called Tim Hornby who was a steward and was working towards becoming an airline pilot. He was also an instructor and I asked him to check me out in the familiar Cherokee, which he did, on a flight I logged as Luton to Luton. That meant I could fly the Cherokee to my heart's delight. Another bonus was that I could fly it at a very good hourly rate because the airline ran a flying club.

So off I'd go, sometimes over to Le Touquet, and re-engaging with light aviation was a delight. It was great fun taxiing out of Luton in this little Cherokee between two huge jets. That got me back to light aviation, because I hadn't flown light aircraft since my flying instructor days. The club aircraft, particularly the faster Mooney M20J, was used by the engineers. I was once grounded in Newcastle with a faulty part and the Mooney meant the engineers could bring up a replacement in an hour, instead of five hours in a van. I told the passengers: 'In a little while, a small blue aeroplane will park beside us and an engineer will bring us a spare part. Then we'll be off to Orlando.' This saved a long delay as we still had enough flight duty hours to operate the flight.

A very interesting study was carried out by British Airways, which discovered a fact about pilots that most of us were unaware of. They found that almost all of their pilots were first-born – like me. I guess being the first-born must develop certain personality aspects that are useful in the profession. Companies need pilots, especially captains, who will conform to managements diktats yet be confident and, if necessary, autocratic enough to run the show. People who are punchy enough to be in charge at the other end of the world, sorting out myriad problems without ruining everything and upsetting customers.

At Britannia, our failure rate when trying to recruit men and women of this rare breed was colossal - 70% on the command course at one point. When I joined, the Britannia command course was renowned in the industry, and that went back to the loss of a Bristol Britannia in Ljubljana in 1966 when a young commander was found to have lost situational awareness. There was apparently a mis-setting of the pressure altimeters because we used to land with the QFE set, so the altimeter would read zero on touchdown. We stopped using QFE because some countries don't have such a procedure, so QNH is now set on the pressure altimeters, which means that on touchdown they read airfield altitude above mean sea level. There is high ground around Ljubljana and there was poor visibility. They hit the hill. Not everyone died - one person who survived flew with us not long ago - but the young captain and most on board were killed.

Don Tanton, who had come into the airline with Jamie in the early days, became chief training captain and his role was to knock this typical post-war charter airline into shape. Charter airlines had a poor name in aviation in the 50s and 60s and his task was to make the operation world-class, which he eventually did through unbelievable discipline and by beating heads together. His maxim was simple: 'You do it this way and you do it every time.' The calls were absolute precision. In Virgin, for example, when climbing with 1,000 feet to go, you might say '1,000 to go' or 'One below', and as long as you said something, that was good. In Britannia, you had to say: '1,000 below.' Anything else was unacceptable.

In the early days of the command course, around 1977/78, the company was growing very quickly. People were coming in and being promoted within a year. In those days, and when I was promoted, you needed 5,000 hours to be a captain, but it is now 3,000 hours. A lot of RAF chaps didn't have 5,000 hours because in the military you tend not to build hours quickly, especially flying fast jets. Ex-military pilots could start the course with 3,000 hours. They would join and within a year they would maybe have a go at a command course. Some would fail simply because they hadn't yet got a handle on the culture within a civil airline, which is understandable. They would havc another go after a year and most would be fine. They stopped taking direct-entry captains in the early 1970s because the people who were arriving were a very mixed bag, from very different cultures. Instead they used ex-military or people like me, from smaller airlines or with no airline experience after completing an ab initio course, and who then built up their hours in the jet. But the command failure rate was high. To be put forward for a command, you had to have an ATPL (airline transport pilot's licence). You needed enough ticks on the simulator tests - at the end of two days, the examiner who had been watching you would tick either 'suitable' or 'not suitable' for command, based on what he had seen you demonstrate by way of technical knowledge and flying ability. If the CSB (command selection board) thought a pilot was suitable, they would start the command course in seniority order. You needed to have the seniority, plus all of those proven abilities, and if the CSB felt you didn't have that, you would be passed over.

If you got on the command course and failed, and the CSB felt you would never get up to command standard, you used to go on to a different pay scale where you got another three increments. You were paid more than a senior first officer to acknowledge that you had a lot of experience, although you were not going to be a captain. In BA, on the other hand, my brother told me that they do the opposite. If you are not capable of being a captain, they freeze your ability to move up the pay scale, and that effectively punishes you for not achieving a command.

Once Britannia decided you could go on a command course, it would start with a few days in the classroom covering all the technical stuff, how to deal with passengers and the importance of not carrying extra fuel without a good reason. Next was the simulator because you had to be cleared in the left-hand seat, and then you'd go on to the line. You'd be in the left-hand seat with a training captain in the right-hand seat, and that continued for about 20 sectors. Then you'd go back into the simulator and operate a command route exercise. This sim check, called CT4, wasn't so much about stick and rudder skills. They would put you in a position where you had minimum fuel and you were going into, say, Geneva in poor weather with passengers. The examiner would introduce a few problems to up the ante and you'd be expected to sort it out. That used to be a real fail point for many pilots. You would have a normal line co-pilot in the right-hand seat in the simulator - the first time, as a potential captain, that you would fly with a regular co-pilot who was just rostered at random for these simulator duties. All these problems would be thrown at you. They weren't trying to break you, but they did want to ensure that you could handle everything successfully. Giving up was not an option.

You also had to manage the co-pilot, as the training captain would be watching your overall performance and thinking: does this person look like an aircraft commander? There were two two-hour sim details over two days, one concentrating on the flying and aircraft handling, and one on crew resource management (CRM) subjects, such as decision-making, task management and communication. The CRM detail used to catch a lot of people out. It was the major fail point of the whole course. Then it was back on the line with a check captain, who was one of a selected few TREs (type rating examiners), so command standards could be maintained. In this final phase, you would spend about five days flying the line with passengers, staying in hotels and going out of different airports, and the check captain would fly with you. He would act as a competent co-pilot but not be at all proactive and would just let you deal with whatever happens. At the end, he would decide if you would pass or not.

My first left-hand seat flight was January 9th 1989 to Shannon, because in those days we also had to do touch and goes with no passengers. Then on January 10th, I flew to Monastir in Tunisia. I was back on line for the final five CT5 flights between February 22nd and 28th, and it was at that time that Dad died. ATC called us when airborne and said: 'Phone your company on landing.' The company told me to go home, leave the command course and start again later, but I wanted to finish my course. They said: 'We can't make allowances for the fact that your Dad's died.' I said: 'That's fine.' I thought: 'Just let me get this done and then I'll go and bury my Dad.' He would have expected nothing else.

On February 28th I ended up in Germany in Dusseldorf on my last hop. I was working out the fuel requirement for the way back and the examiner said 'I'll fly it back', which either meant I had passed or I had failed – I wasn't sure which. As he walked out of the flight deck into the cabin, the examiner said to the crew: 'Come and meet Britannia's newest captain.' I had passed my command course. I thought: 'I'm a captain and Dad never knew, but he would be proud.' I didn't do anything to celebrate. I didn't really feel like it.

Since my Mum's death, Dad had slowly become little more than a passenger in life. Within two years of her dying, he had come to Olney to live permanently with us. The deal was that I was going to sell the family house at Giffnock and use most of that money to find a bigger house in which Dad could live with us. Ruth and I bought a plot of land for £60,000 and it took about two years before our new house was completed. We finally moved in in December 1988. To mark the occasion, a picture was taken of me in my first officer's uniform, because the agent wanted the photo for publicity. I thought: 'I'd better get this command now, otherwise we will be seriously skint.' A few weeks later, just before Christmas, Dad moved in with us, about three months before he died. By this time, my brother Nigel had started flying with BA Highland Division, based in Glasgow.

Getting on top of ETOPS

In 1987, Britannia had announced that we were going to go long haul for the first time in many years. Up to that point, the longest flights we were doing were about four-and-a-half hours to places like Cyprus and Tenerife with the occasional Banjul, Gambia. In early 1988, I became a training first officer on the B767, and I was told about this new thing called ETOPS (although it was called EROPS initially) which the Americans were going to start. 'ETOPS? What does that stand for?' They replied: 'Extended Twin Ops.' It was introduced to allow two-engined aircraft to fly more than 400 nautical miles away from an airport. We wanted to be one of the first European airlines to fly ETOPS because we were going to start long haul to Australia and Florida. We could have gone to Australia non-ETOPS but the route would have been longer and with fewer options. ETOPS was required for the long flights over the Arabian Sea and the Bay of Bengal, as well as parts of Australia. They asked if I would write the ETOPS course. I knew absolutely zip about the whole thing. I don't know why they asked me, but I had the summer to do it.

'Where the hell do I start?' My first plan was to go to the United States to fly with Piedmont, an American airline from the south, based in Greensboro, North Carolina. I flew a couple of Atlantic crossings with them, as a third pilot in a B767, because they had already started ETOPS flying in the US. I spent a few days looking at their course and what was covered. The ETOPS procedures are based on the ICAO (International Civil Aviation Organisation) rules. They cover pilot training and procedures, which was the section I was producing, and they also covered engineering procedures and extra aircraft equipment. For the airline to be certified for ETOPS, all the different aspects had to be approved by the CAA. I was also writing a separate module on the North Atlantic operation.

To cross the Atlantic, most aircraft fly on the NATs (North Atlantic tracks) and it is unlike anywhere else on the planet. The routes (tracks) are created twice a day to take account of the shifting winds at altitude by either Shanwick - a combination of Shannon and Prestwick - or Gander, depending on direction. One does day and one does night. They are based on the shortest flight time between New York and London. In the day, the tracks are A, B, C, D, E – A being the most northerly track – and at night, it's from Zulu going back up north. There are usually about five tracks, sometimes more, and they are 60 nautical miles apart, ie one degree of latitude. By day, they start to the west of Ireland and they run across to Newfoundland. There might be another track lower down, from about Lisbon across to Brazil, and as that would be the most southerly, it might be Track F. Anything else is called a random route.

So you are either on a NAT track or a random route and they start from 29,000 feet up to 41,000 feet. Above that, you are on a random route, and if you are lower, you are on a random route. If you started on a NAT track but at a point along the track your route took you off it, that would still be a random route. The NAT tracks are published in the morning for the day flights, so when you get to the crew room and download the OFP (operational flight plan) along with the weather and NOTAMs (notices to airmen), there will be the NAT track message with that day's tracks, set out in lat and long, and available flight levels. To the Caribbean, it is usually a random route and to Orlando a NAT track. It is only when you are at top of climb and in contact with Shanwick on VHF radio that you get your actual clearance for the Atlantic crossing. The first waypoint on the tracks will have a name, so they are fixed points. Thereafter, we use latitude and longitude for the rest of the waypoints until we reach the exit waypoints, which again are fixed points. The first lat and long going west might be 53N 20W, then on to 54N 30W, where we contact Gander on HF Radio. At night, the procedure is reversed.

There was an official ICAO document about the Atlantic, describing the NAT tracks which had been designed after the war for the early jet age. You could cross as far up as Greenland on a track or down to the Azores, depending on the weather over the Atlantic. A flight to Florida could mean flying across one day on a northerly track to keep us out of the strongest headwinds. Wind aloft always comes from the west, so you have headwinds going and tail winds coming home. The return flight overnight would be on a more southerly route to get the tailwinds. The tailwind on that route might only average out at about 50 knots but over seven or eight hours, that makes a big difference to flight time and therefore fuel burn.

The lowest you'll find the jet stream over the North Atlantic, is about 28,000 feet. Sometimes it will go up to the high 38,000s, but then you are going into the stratosphere. The tropopause is the layer between the stratosphere and the troposphere, where we live. The tropopause varies from about 30,000 feet at the poles, up to about 56,000 feet at the Equator. The heat at the Equator and the height of the tropopause mean lots of room for the huge thunderstorms you get in the Tropics. Because the troposphere is thinner at the poles, when operating on polar routes the cosmic radiation in the cabin can cause leukaemia. The North Pole is where particles from the sun and galaxy are sucked into the atmosphere by the earth's magnetic field. Therefore, there is much less radiation at the lower latitudes.

If an airline regularly flies over the polar region, there are guidelines about flying when pregnant. Concorde used to have radiation sensors because of the height it flew at. A warning system was fitted, but I believe it was never activated because they weren't up there very long. Britannia had a trial, and BALPA and the pilots accepted the results. It was never a major issue. Of course, where radiation could become an issue is over the length of a career. The big change was around 2007, when internationally the pilot age limit went from 60 to 65. It had been on the cards for a long time but the Americans were against it. The big carriers were going bust and some of the pilots' pensions were destroyed. Once the American authorities agreed, ICAO changed the retirement age to 65. After the war, people had been drinkers and smokers, but as the years went by, pilots were living longer and healthier lives. Radiation was never a factor in that.

Aeroplane windscreens are about an inch thick and they are built up of layers including gold, which is heated with an electrical current so the whole screen is heated. You can have a complete smash from the outside, and the windscreen will stay intact. Pilots are advised to wear sunglasses with brown lenses because of the ultra violet rays at height - if you wear blue sunglasses, they do not protect your eyes from the damaging rays from the sun. My eyes now are pretty poor but I don't know if they were always going to be poor. Even on the nastiest day in November, you take off and a few minutes later you are in the sun and spend the rest of the flight in it. Back in the B737 days, before we did this radiation test, some of the skippers might say: 'I'm not going any higher than… such and such height' because the difference in radiation between 30,000 and 40,000 feet is colossal. The B737 couldn't get that high but it could get to 35,000 feet… on a good day, 37,000. In the end, the radiation check was done and it was found to be okay.

At the time I flew with Piedmont, they were very charming and helpful, but it was the first time my beard came into question. At the end of my visit, I went to find the boss to thank him, and he asked why I wore a beard as a pilot. In America, you have to be clean-shaven, as they reckon the oxygen mask won't seal very well on top of a beard. In the UK, you can't have a beard in the Army, you can have a moustache in the RAF, and in the Navy you can have a beard. But that's not part of European rules. The oxygen we have, unlike the passengers, is pure oxygen, 100 per cent and at pressure, so it is pumped into us and we only have to breathe in deeply.

I spent some time at Piedmont looking at their courses because ETOPS was brand new. ETOPS was initially introduced at two hours, which meant that, at the planning stage, the flight had to remain within 120 minutes of a suitable airport. The first thing they did was come up with a high-speed single-engine descent and diversion, so that would allow you to fly further from an airport. The 120 minutes translated into 820 nautical miles, which meant you could never be further than 820 nautical miles away from a suitable airport. The silly thing is that you could fly over the whole of Europe for four hours, with all the airports covered in fog, and that would be okay because it isn't ETOPS. You just need to be no further than 400 miles away from an airport, open but unusable in thick fog. Rules is rules!

At the planning stage, the route from Gatwick to Orlando would usually take you out over Ireland and, at some point, during the initial cruise, the pilot monitoring would go off the air and on to the other VHF radio and call Shanwick for the oceanic clearance. Shanwick would require an ETA for the expected oceanic boundary waypoint, our requested flight level and Mach number, then we would wait for up to five minutes and they would call us back with the clearance. Fingers crossed, it would be the one you had already loaded into the flight management computers. They would say: 'Ready to copy clearance.' The other pilot would listen in as well, so we would be working two frequencies. Like all things, it is reasonably straightforward once you are used to it, but when I was reading about it for the first time, I thought: 'This looks terribly complicated.' So I went up to Prestwick, where the Shanwick controllers are.

The people you speak to on HF radio are radio operators, not controllers.

Over the ocean, you call up with a position report, for example saying: 'Shanwick radio, this is Thomson 123 position...' They will say 'Go ahead' and you give your position in a formal format: 'Shanwick, Thomson 123, position 55 North, 20 West at 10:04, Flight Level 330, estimating 54 North, 30 West at 10:41, next 54 North, 40 West.' They then relay that information from Shannon where they are, to Prestwick, where the controllers are, and they make the decisions. That comes back to the radio man, who then sends it to the pilots.

The whole process is pretty slow, but that is the way it is. And first of all, you have to get through to them. VHF is line of sight - at altitude you have a range of about 200 nautical miles, so fairly quickly we changed to HF radio for the crossing. The HF signal bounces off the ionosphere and I have been over the Atlantic unable to speak to Shanwick or Gander. All I would be getting is crackle - and then I would be hearing Perth in Australia clear as a bell. I once spoke to Perth and they said: 'Where are you?' I replied: 'I'm over the Atlantic!' It is really pathetic, especially when there are passengers sitting back there using their satellite phones with no problems. Now there are sat phones fitted to the aeroplanes and also CPDLC (controller pilot data link control), and position reports or requests, eg. for climb or en-route weather diversion, can be sent through the flight management computers, so there is less time spent on the HF radio.

We used to have to plot our route on a plotting chart. We would also plot the NAT track either side, if there was one, and 10 minutes after passing a waypoint, plot our position from the computer to make sure we were on track. This is no longer a requirement because GPS is fitted to all the aircraft. If you were on NAT track Bravo, that would mean there was Alpha to the north and Charlie usually 60 nautical miles to the south, and all three would be plotted on the chart. It was a useful visual reference of where you were. If you had to turn, in an emergency due to an engine failure, you would turn off the current track since you could not keep altitude on one engine. You would turn and descend as slowly as possible until 15 nautical miles clear of the track, then accelerate and dive down to get below the track system and turn for the alternate airport.

When you received your clearance and you were on a NAT track for the whole crossing, you acknowledged the track letter and the TMI (track message identification) number, plus flight level and Mach number. The TMI number is actually the day of the year. So on the 5th of January, the TMI would be 05. If you were on a random route which was not on a track, you would read the whole thing: 'Thomson 123, cleared Orlando via Amoco, 52 North, 20 West, 52 North, 30 West, 50 North, 40 West, 46 North, 50 West, 38 North, 60 West, Flight Level 340, Mach .79, maintain FL 340 and Mach .79 from Amoco – read back.' If it was a re-route, you would have to plot the whole thing again. You would also plot equal time points, the point where it is the same flight time to the current two diversion airports.

If you wanted to change level at any time, you would call up with a request, but it all took time. They always know when there is a more urgent request. If you get an emergency, you can declare a Mayday. All aircraft listen out on 121.5, which is the international distress VHF frequency. If you were having a major incident, you could call on that and someone would start relaying for you. The bottom line is if you lose an engine, you can only hold height for 30 seconds, depending on how heavy you are, so you are going to have to start coming down. Procedure if you lose an engine is select limit thrust on the live engine, trim the aircraft with rudder, and as the speed would be decaying, turn either left or right by at least 45 degrees. If you are going west and your suitable on route alternate is to your left, that means you are going to turn left. You are trying to hold on to as much height as you can, drifting down gently so you are out of the way of the track. Once you are clear of the track, you would dive down to 28,500 feet, below all of the tracks, and turn towards your diversion airport.

At the planning stage, to keep within the ETOPS rules, they planned you at high speed to your alternate airport. On the day there is no actual requirement to land within the two hours, it is just a planning exercise. You must, however, remain at all times within the rule distance. The distance figure matters because wind speed and direction affect how far you can fly in two hours, which can be more than 820 miles in theory. After a few years, 180 minutes and 1,200 nautical miles was introduced. So ETOPS, with all its rules and procedures, is how we operate over remote parts of the world. However, it is now almost a non-event. The B777 was the first aircraft to enter into service with 180 minutes ETOPS certification. The A330 was granted 240 minutes and the B777 can be extended to 330 minutes. Airbus is asking for 420 minutes. With ETOPS of seven hours, an aircraft can reach any airport in the world on the remaining engine. A long, long way on one engine.

I have never lost an engine over the Atlantic but I have lost the centre hydraulic system, which is the main system, on the B767 when I was about 30 West going to Orlando. I had to decide whether to turn back or keep going. There was a lot of equipment which was driven by the centre hydraulic system that I didn't need until I came in to land, including the wheels and the flaps, so I thought: 'I might as well take it to Orlando. We have an engineer there and they can fix it there. At least the passengers will be pleased, and I have a shopping list!'

I used to do a thing when I was passing Cape Hatteras on the North Carolina coast where I would say to the passengers: 'Here is a quiz and the winner will be able to get into the flight deck for the landing. What happened at Cape Hatteras in 1903? And it could have happened inside the length of this aeroplane.' Of course, it was the first flight by the Wright Brothers. No one ever got the answer but the person closest would be let into the flight deck anyway. You can't do that these days.

I thought I should tell the American controller something about my hydraulics failure, so I said: 'I'm not declaring an emergency, but I have lost all the fluid in my central hydraulic system.' From the UK to Atlantic City is just about a straight line but after that you have to fly right of a straight track due to military airspace off the coast. I advised ATC: 'I am continuing to destination.' They said: 'We'll call Orlando and get them geed up and you, sir, are cleared direct to Orlando.' I was amazed: in one stroke I was cleared through all the danger areas the whole way down the coast.

During the descent into Orlando, we were working our way through the quick reference handbook. We had to lower the landing gear using the alternate procedure and run the limited flaps out electronically, thus giving us a higher landing speed due to the limited flap available. I said to the passengers: 'You will no doubt find we are surrounded by fire engines on the ground but don't worry about it.' As I started the approach, the controller called another aircraft saying: 'We have a heavy B767 with an emergency and we are going to give you a few more track miles to extend the gap.' I thought ooh that's me! In the UK, you only say 'heavy' on the first call on each frequency. This is to remind the controller that they have to give a gap of a few minutes, depending on the size of the next aircraft, due to our wake turbulence. In America, you use 'heavy' on every transmission.

We landed, and all was well. It was the first time I had landed the B767 with Flap 20 in the real aeroplane. On selecting the gear down on the alternate system, the two main gear – wheels - dropped normally, but there was no indication for quite a while that the nose gear was down and locked. During this time my mind went into overdrive as I thought: 'Now we are doing a partial gear-up landing.' After what seemed an age, we got three green lights. If we had an indication that one of the gear was not down, we could request a low flyby of the tower for them to have a look and advise if it was down. We, as an airline, felt that flying down the runway low and slow to see the wheels doesn't really tell you very much at all. The controller would only be able to tell you 'I can see your wheels', but they couldn't tell you they were down and locked. This just upsets the passengers anyway, so we tended not to do that as a normal procedure in this non-normal situation. You end up trusting the sensors.

The fire engines followed us down the runway and, once we stopped clear of it, they surrounded us. I made contact through ATC and the fire chief said all was well and they would follow us in to the gate. After shutdown, when the crew opened L1 door (forward left), there was this big American fire chief standing on the air bridge all dressed for action and he said: 'Welcome to Orlando, Cap!' I said it was very good of him but it wasn't necessary, to which he replied: 'Ah gee, Captain – best being at the runway, no point in being in the shed.' I said: 'Thank you, you are absolutely right.' And that was my first real in-flight problem after I completed my command course.

Learning to work my authority

I had to present the finished ETOPS course to the training captains. The day came and I had everything on slides, using the old overhead projector. Steve McKinnell, another training first officer, was going to take over from me because I was leaving the fleet. The training captains would be taught the ETOPS rules and procedures and they would, in turn, teach the pilots. The plan was to do a day's ground school on ETOPS for the pilots who would be flying long haul, followed by a written exam. At some point, the trainers were going to have to sort it out because someone had to fly the jet for the first time.

When I wrote the course, I had been in contact with Boeing a lot to understand about our aeroplane – the Britannia ETOPS B767-200 – which was being built in Seattle. I had to know, for example, what size fuel tanks it had. You don't buy an aeroplane like you buy a car. You buy a Boeing, which then has a six-inch book of possible specifications, including fuel tanks and engines (General Electric, Pratt & Whitney, or Rolls-Royce). You could have cross bar flight directors, V bar flight directors – there were a million different options, which Boeing has since stopped.

I spoke to our engineer – we always had our own Britannia engineer at the build to make sure that what went in was what we wanted - and asked him: 'What size tanks have we got?' You could get 56-tonne tanks, a 65-tonne tank, and a 76-tonne tank. Part of the technical information I was including in the course was the new ETOPS specification and the size of the fuel tanks - and I kept getting different answers. At one point, it was 76 tonnes and then it was 65 tonnes, and when I eventually got to Boeing, we were actually standing underneath our jet and no one could tell us what size tanks it had. So they filled them up, and found they were the 56-tonne tanks. The 70-tonne tanks would have been for the Pacific runs and if we had those fitted, it would mean we would have to carry less load. You can only carry so much weight, and we could never carry 70 tonnes of fuel and a full load of passengers with bags, and possibly some freight.

I asked if I could take some leave because I had never really seen America, and they said: 'Take a few days. Just be in Seattle on the right day.' I suggested Ruth and I fly to LA, which they agreed. Our friends Karen and Camille were to be in LA visiting Cami's relatives, and we told them we were flying out. We flew to LAX direct and stayed at The Registry, a very posh hotel where the vice president was staying the night, so there was security all over the place. To a tourist like me, I was impressed by the big cars, the cheap fuel, and the food. But scratch the surface and I soon discovered that living there would be a different matter. Camille and Karen showed us around LA, then Ruth and I hired a car, drove north along the coastal highway and spent the night at Big Sur, just south of San Francisco. We spent a day in San Francisco and flew to Seattle in the evening. The chap at the ticket counter couldn't give us standby tickets, even with the letter of introduction I had been given, but I guess he felt sorry for us: he issued Ruth a very cheap ticket that should have been purchased months before, and he gave me a crew standby ticket.

When we arrived in Seattle, Chris Norman, our chief pilot, met us at the hotel, and one of our engineers was there. Chris said we were going to be there for a few days because the press and television people were not ready for us yet back in Luton. On one of the days we were shown around the huge Boeing factory, where they built the B767s and B747s. The B757 was built in another factory at another airfield where the B737 was built. This factory was the biggest building, in terms of volume, in the world, and inside there was a line of B747s and another line of B767s. They even had their own nuclear bunker, which they were very proud of. This building was so big, they said there could be weather inside it.

When we departed for Luton, there were only a few of us on board. Ruth was sitting in the flight deck for departure, and in the back there were a few engineers - and an accountant to make sure all the finances were in order. The empty aeroplane had been painted in the new Britannia colours and it looked fabulous. This was important for Boeing because it was the first European ETOPS B767. Britannia had always been a very Boeing company, so we were an important customer. We were going to go direct, Seattle to Luton, which was fairly straightforward weight-wise. Because we were going out of Seattle over on the west coast, we were taking a very northerly route, high up over northern Canada and the middle of Greenland. Neither of us had ever flown that route. We were not into polar navigation, but we were far up north. By the time we reached the Atlantic, we were well north of all the NAT tracks on a random route. Of course, no one in the company had flown ETOPS so this was completely new. The CAA had authorised Chris Norman to operate as commander because he was a training captain and the boss – someone had to do it. He had flown a B767 around Europe but never transatlantic. I think he regarded it as a bit of a management perk.

We took off in this empty aircraft in the evening and by the time we got up to northern Canada, the Aurora Borealis was magnificent. We had Ruth and the chief pilot's wife Helen in the flight deck and everyone else was down on their knees looking out of a side window because there were only a couple of rows of seats in the whole aircraft. I knew about the Aurora because I had seen it years earlier in Narsarsuaq with Davie when we flew the DC-3 back from America. On this occasion, it was as if we were an ocean liner on a moonlit calm sea, and sailing between colossal icebergs on either side, lit up in the moonlight. It was magnificent and breathtaking. In all the years since then, I have never seen the like again.

When we came on to the approach into Luton, early in the morning and with Chris flying the aircraft, we decided we were going to check the autoland. There are three autopilots. Nowadays, it's automatic as you set up the instrument approach, but then you had to select the other two when on final approach. And Chris had forgotten. I said to him, so as not to make a big deal of it: 'I'll select the other two autopilots.' What I learned from that was that all people can make mistakes but with him, there was no embarrassment. He was just grateful to have it pointed out. The fact was, the team was working. He didn't make an excuse. He was quite happy for me to remind him, and I learned from that. There are pilots who, when you tell them they have forgotten something, mumble: 'Oh yes, I was going to do that…' Or they are cross with themselves for forgetting. He was neither of those things. He had great confidence and humility, a rare combination.

We landed back in Luton and met the media. Remember that BEA was still flying Tridents at this time, and Dan Air still had Comets! We were way ahead. I once landed a B767-200 in Paphos and parked it beside a Russian Tupolev TU-154 three-engined passenger jet. These Russian guys came over, out of uniform. In fact, they had towels to go swimming and they were preparing to jump over a fence to go to the beach close by the airport. They said: 'Are there only two of you flying this?' Of course, the answer was yes. We told them what the fuel burn was and what load we carried. They then took us over to their aeroplane, which looked like a Wellington bomber. Their flight deck was huge, and there was a navigator and an engineer. This was in the mid-80s and they were still flying this coal-driven jet.

The flight back from Seattle was in November and I flew the aeroplane once after that, on November 24th to Malaga, and that was my last flight on the B767 as a first officer. I started my command ground school in December, and on March 12th 1989 I had my first flight in command. It was on one of the two B737-200s which were occasionally used as freighters. They were big enough to stick in a B767 engine, broken down into sections, if required. After that, I flew to Turin with passengers, a late ski flight. As we taxied out, we were given a re-route just as we reached the runway and were cleared for take-off. Even though pilots and ATC work closely with each other on a daily basis, they sometimes do not appreciate our periods of high workload. As pilots, we occasionally visit them to get a better understanding of what they do and in the past they would join us on the flight deck to see us in operation. Despite this, we still find it difficult to get across to them certain basics.

One is they shouldn't try to give us instructions when we are decelerating on the runway and still in reverse thrust. We are too busy landing the jet to take on board the information they are trying to tell us. We were literally coming on to the Turin runway when they made this announcement. I was pilot flying and they rattled through this re-route and expected me just to get airborne. On a B767 you could tap this information fairly quickly into the flight management computer and that would tell you your predicted fuel on arrival at destination using the new route. On the old B737 we had no such computer. We would need to get the maps out, find the new waypoints and establish the track miles of the new route, determine if it was longer and therefore how much more fuel we would require. I thought: 'I don't recognise the new waypoints and I can't just launch and hope I have enough fuel. I certainly can't just sit on the runway and get the route maps out.' There and then, I made my first good command decision. I said: 'I am not accepting that.' I taxied back off the runway - and worked out that we needed another tonne of fuel to follow the new route. I had to refuel and explain all this to the passengers.

After that little hiccup, my flights continued in the B737 around

Europe. I was finding it much easier being a captain than a co-pilot. I enjoyed working with the crew and running the show. I have always felt it was an easier job because as a first officer you might be about to take a sip of coffee and the captain will say: 'Come on, hurry up. We've got a slot to make.' I like to focus my briefings to the first officer on what we are planning to do and also to cover the different points if something goes wrong. For example, what I am going to say on the PA to the passengers if we have to evacuate – 'Evacuate, evacuate. Undo your seatbelts and get out!' This evacuation procedure was introduced in the late 1980s. I'd say to the first officer in my briefing 'Evacuate, evacuate. Undo your seatbelts and f*** off!' in the hope that generated a wee smile and I had their attention.

We can end up briefing each other to death, so I just liked to shake it up a bit. I know the chart date is important so we are looking at the same charts, but if you just read through them, the other pilot will switch off. Concentrate on what is going to happen, use eye contact as much as possible, and update the brief as you go along. I'd say: 'Once I have said that and they have evacuated, jump out of your seat and look through the spy hole on the flight deck door if we haven't heard the main door evacuation slides deploying. But don't open the door to the flight deck or they will all be in here! Make sure they are not all still sitting there if I have pushed the wrong button for the evacuation announcement – because that is all it would take.' There is a row of buttons and I could have pushed the wrong one, thinking I was still in yesterday's aircraft, because they are in a different order on some aeroplanes.

I was flying out of the usual UK airports, but still Luton-based, and it was a busy summer. A flight from Luton to Paphos would take 4 hours 15, but we could seldom make Paphos direct to the northern UK airports on the B737-200, so we would have to stop at Ostende or somewhere en route to pick up fuel. It would be up to me where we stopped. We would just push it as far as we could. We were an established, well-financed company and known, so we could pick up fuel pretty much anywhere. The captains all had company cards for that sort of thing.

I was 34 and a young captain. Quite a few of the first officers I was flying with were older than me. We would position not in uniform somewhere, maybe go up to Glasgow to fly the next day, and when we arrived at the hotel, they would say to the first officer: 'Hello, Captain.' He would have to say: 'Well, no, actually – he is.' I quickly learned to arrive at the check-in desk first. I have always taught that when you fly with people, training captains or whoever, none of them are perfect.

What you have to do is to pick out the good bits of their operation and leave the bad bits behind. You make up your own way of operating and running the show within all the rules and standard operating procedures. I had done that while I was completing my apprenticeship as a co-pilot and I knew that I didn't want to bully people. I wanted to let the first officers have the opportunity to fly the aircraft the way they wanted to fly it, without me being all over them – unless, of course, I felt I was losing my comfort blanket.

I was learning how to work my authority, yet still run a fairly casual flight deck – and a flight deck open to good ideas from any of the crew. I was learning how to look after the crew and care about their well-being. I think I was a laidback, easy-going captain but we always flew to a high standard. When I joined the airline, there were very few stewards in Britannia. By the time I got my command, stewards were normal accepted crew. In the earlier years, some of the captains didn't like that 'their girls' were not always girls.

I landed into Edinburgh once when I was a first officer and the airport had stayed open specially for us because, in those days, Edinburgh shut at night. We landed and parked, the passengers were deplaning and we were getting our things together. There was a bit of a to-do going on outside because the new crew were walking out to the jet and the captain was pushing a trolley with his luggage on it. The to-do was because a baggage handler told the captain he couldn't bring his trolley out. He had replied: 'Don't talk to me, my man. I am the captain.' So they all said to him 'Stuff you' and walked off the jet, muttering: 'F****** pilots.' I said to my captain I was going to see if I could sort it out. I went up to the guys and said: 'I am sorry about this. I used to be a baggage handler and I know some of the pilots are arses, and here is as big an arse as you are going to find.' They came back to unload and then load the jet. They were quite right to have told the oncoming captain not to bring the trolley out on to the apron because that was the local rule. The vast majority of captains would have said: 'Oh sorry, I didn't know. Where shall I put it?' He was just not going to take it from a baggage handler. I have seen that same man make one of our female handling agents cry on the flight deck over some trivial issue. The word bully springs to mind.

Managing a crew of my own meant dealing with all sorts of situations. A girl once came to my room and knocked on the door - as you can imagine, there was a fair bit of door-knocking going on. I opened the door and this girl was in tears. What's the matter?' She said: 'I'm in love with the first officer.' Come in, I said, I've got his room number. She announced: 'No, I've just been there and he's in bed with the steward.' I sat on the bed with this inconsolable girl and thought: 'Good grief, is this part of a captain's life?'

The real skill of managing a fixed crew – who are together for a series of flights as opposed to changing daily, or even after one sector – began when we started to go to Australia and were away for three or four weeks. There were still the usual parties, with the crews sleeping with whoever. I was still expected to go out to dinner with them and spend time listening to their outrageous stories, which I usually enjoyed as long as I had a good crew. The first sign of trouble was the dinner bill on the first night. If that went smoothly, the chances were that the rest of the trip would be fine. I had to make up the bill many times, with crew members saying: 'Well, I only had a salad and one drink.' A good crew will usually suggest before I do: 'Shall we have separate tabs or split the bill?' I would sit with the crew and think: 'I couldn't tell many people about some of the stuff said at this table.'

Of course, for younger cabin crew members, off-duty is their play time. They work hard and they fly more than the pilots do because of our hour restrictions. At home, they might share a flat, and on their few days off they would be cleaning, washing and ironing clothes, and getting ready for the next five-day trip. When you are away on a long-haul trip, it is the crew's play time, socialising away from home, kids, family, partners, bills and problems.

I was approaching my mid-30s, I was a father of three and I had been listening to the same sort of stuff from the same kind of people for 10 years. I wasn't yet at the point where I felt my ears could not take any more – but that point did arrive. Frankly, there was no orifice that I hadn't heard about. Just when you think you've heard everything, someone will surprise you with something outrageous. There is nothing… nothing… that I have not heard!

Blowing in the wind

February 8th 1990 is a date that will live long in my memory because that was the day of a major storm, with 80 knot (96 mph) winds across parts of the UK. On the way back from Malaga, we were coming into Stansted, over the Channel Islands. I looked down on the Channel during the descent and it was white. To get white water, you need about 20 to 25 knots of wind. I thought: 'It is quite windy down there and that's not forecast.' Approaching the Isle of Wight, I could hear on the RT that aircraft were diverting, so I knew something was up. We crossed the coast at about 25,000 feet descending and I could hear all this going on but we were carrying on and listening to the various Met broadcasts. We got to Stansted and it was blowing fairly strongly. The ground speed was slow and it was very windy, blowing from the south west. I held for a while until I was cleared for the approach, and it was only during my approach that there was a warning from the tower. ATC told us to go round - and then they announced: 'We are evacuating the tower. The glass is coming in!'

I headed for East Midlands because the weather was better there. Because I had held over Stansted, I didn't have a lot of fuel. On handover to East Midlands, ATC started radar vectoring us for an instrument landing system approach to the westerly runway. I was number one for the approach and the wind was blowing from the west at about 60 or 70 knots, almost down the runway. I knew we were short of fuel and there was nowhere else to go. I briefed my first officer: 'If we don't get in on this approach, we've only got enough to come around and land. And if we don't get in because the runway is blocked, we will need to land on the grass. There is no Plan B.' Suddenly, we had gone from enjoying a normal day of flying to 'somehow we've got to get this jet on the ground'. My biggest concern was that we had to come up with a plan that would give a good outcome within the time available, dictated by our remaining fuel. I knew we couldn't divert to Birmingham because the runway direction there meant the wind was straight across, and the grass at East Midlands was a better option. So we had to land at East Midlands, irrespective of the runway being blocked or not. I knew only too well that, in that sort of weather, someone could pop a tyre, or something else could go wrong very easily.

We landed in one piece and taxied in, but then discovered that we couldn't get our passengers off. East Midlands doesn't have air jetties and as the steps came over, they were blown over. The old B737-200 had air stairs – its own stairs that came out from under the floor at the front left door – but they had a 40-knot limit for operating. We sat there, rocking backwards and forwards in the gale, and ended up on board for hours. Of course, the passengers were getting hacked off, but I couldn't do anything because the steps were getting blown over. Eventually, the wind died down, we disembarked and stayed the night in the hotel at East Midlands. By the time I got to my room, the TV news was reporting that this storm had flattened almost everything in its path, but these conditions had not been forecast. I thought of the Stansted control tower glass blowing in and wondered what debris might have been on the runway. At that moment, our situation moved from 'this could be a tricky landing' to 'I'm now flying again, climbing into a storm with limited fuel and this is not good'. I don't think I considered for one moment landing on the runway in front of me, but if I could have had the big picture and known what was ahead, I might have said in reply: 'Negative, I am landing straight ahead.' East Midlands has an east-west runway and the wind was less strong further north at that time, so that was the best runway for us to go to with that wind direction.

What you learn from all that drama is there has to be a successful outcome. In later years, we saw no better demonstration of that than the water landing on the River Hudson in New York. That pilot went for a successful outcome, a realistic outcome. It might have meant the destruction of the aircraft and it might have included people dying, but it was always going to be a better outcome than everyone dying. A similar thing happened to a TriStar which landed in Jeddah with a fire on board, and everyone died because someone didn't take control. It had landed back, so the pilots were alive at that point, and the aeroplane stopped on the runway. It was after that everyone died because no one took charge. You have always got to think these things all the way through to a real outcome. Later on, when I sat as an examiner in a simulator, I could see when people were not able to do that. I debriefed a first officer who had done a detail where he didn't bring the gear up after take-off. There had been a bang and vibration just before lift-off and he thought a tyre, or tyres, might have exploded. He didn't want to retract spinning shredding tyres into the wheel well, since the flying rubber might have damaged hydraulic lines or control cables. He didn't have a lot of fuel as he had gone around from Gatwick on the first approach because there was an aircraft on the runway. He was coming downwind and said they could divert to Heathrow or make another approach into Gatwick. He had a few options – he could have brought the gear up after a time because the wheels would have stopped rotating and cooled down if they had been hot, but there was a chance they could have been damaged and jammed up.

He briefed that he would have another go into Gatwick and if he didn't get in, he would divert to Heathrow. That was the brief he gave to the captain. It was the first officer's sortie and was one of the simulator opportunities to help him develop for command. The captain takes a step back. I wrote down so I could debrief him: 'You must land off this approach into Gatwick. If not, you are going to crash somewhere.' Other than that, he had done very well and I said to him, as we sat in the debriefing room: 'That was pretty good, in general. Captain, tell John something he did well.' I got them chatting together, and I said: 'What's the big thing we have learned today? What can we take back to the jet?'

There's no point in learning these things in the simulator unless they take that learning on to the jet. 'Let's just go back to the point when you were downwind with the gear down. Do you remember how much fuel you had at this point?' No. 'Well, let's say that is 1,200 kilos. And you briefed that if you didn't get in, you would go around and possibly divert to Heathrow.' It was only at that point that the penny dropped that he hadn't really analysed the fuel remaining and the extra fuel-burn with the gear down. He went: 'Ah!' What that debrief needed to include was that there was no go-around - and air traffic needed to know there was no go-around. They needed to be told that the Gatwick runway should be immediately sterilised and no one could use it for take-off or landing until he was on the ground.

He should have declared a 'Mayday' to alert ATC of the urgency of the situation. He hadn't thought about a plan B or a plan C. Plan C might be ugly and include landing on the taxiway or on the grass at Gatwick, but the passengers would have walked away from the wreckage. If I had let him carry on, the engines would have stopped, just run out of fuel, with catastrophic results. He would then certainly have learnt the lesson. To the pilots flying the sim, it is very real. They are flying the jet, as far as they are concerned, and the outcome would have been shocking.

There are very rare situations where ditching in water is one option. Boeing introduced a particular ditching technique, so in the sim we dealt with the more obscure theoretical aspects. As a co-pilot, and even as a captain, I was always advised: 'When you're flying along bored, going down to Tenerife quietly, what you need to be thinking is: what would you do if you had a double engine failure, a decompression, a triple hydraulic failure…?' In that situation, I couldn't make land so what am I going to do? The answer is ditch! I would turn towards the coast to give us the best chance of surviving. I am going to kill people, of course I am, but that is the hardest thing to teach: that sometimes there isn't going to be a good outcome. There are all sorts of rules with fuel management: you need to have this much, you need so much contingency, you need to hold for half an hour, you need to divert… However, when it comes to it on the day, you've got to use the fuel you have. Once you go below two tonnes, you have to put out a 'Pan call'. Once you get a low fuel warning, you have to declare a 'Mayday'.

You could be short of fuel for many reasons – you couldn't lift the weight from the runway, you hadn't considered the extra fuel for diverting around the en route thunderstorms, or perhaps ATC is telling you to hold at a boundary. Whatever the reason, good judgment is required to sort it out. Possibly you should have called in, to for example Lyon, to pick up more fuel on the way home and not carried on with your fingers crossed. There are crews who have screwed up - we all can. We tanker fuel around because it's cheaper than the destination fuel, even in these days of environmental awareness.

Sometimes, either Operations or – more likely – the pilots will make a mistake with the fuel uplift. The idea is to pick up as little of the expensive fuel as possible. When refuelling at the destination, where fuel is expensive, you'll pick up five tonnes instead of ten. On one occasion, the flight crew screwed up and tanked too much fuel, so when they arrived, they were overweight for landing. It turned out that they could lift the fuel load but they couldn't land it. You would have to hold off and burn fuel to get down to the maximum landing weight. Very embarrassing.

When good is not good enough

In the simulator, when trying to give a first officer command practice, it can be difficult for a line captain to let go of command. A training captain would be more comfortable letting the other pilot run the show, even if it's not going well, but some captains just don't, and why should they? They are not actors. In the sim, I would brief that on one detail the first officer will lead the show and tell the captain not to give too many suggestions. The question arises: 'Is this first officer, who is briefing that they will divert to Heathrow after another approach and go-around at Gatwick with only 1,200kg of fuel, ready for command?' The simple answer is no. He'll never do that again but it was interesting to watch his thought process. In the command course, there would be a module on fuel management. There are so many things that can go wrong that you can't cover them all in the simulator.

The leap in technology between the old B737-200 and the B767 and B757 was colossal. They flew higher, faster and further with twice as many passengers, and the introduction of vertical and lateral navigation using triple IRSs (inertial reference systems) featuring triple laser gyros was a fantastic tool, providing situational awareness information to the pilots. It gave the pilots a computer-generated path, or road, in the sky – both laterally and vertically. The autopilot would fly the route and the autothrottle would maintain the correct speed, from just after take-off to touchdown. But as the years went by, sitting back watching the automatics fly the aircraft, a new sinister problem arose. Pilots around the world started to become remote from the jet. The automatics always worked, the engines didn't fail, we flew along the magenta line to the blue circle, the destination, every time. Not much 'what if?' thinking was going on. Aircraft were crashing because pilots were remote, leaving it all to the automatics.

Later, when I was chairman of the command selection board, I introduced the principle that if someone failed a command course, one of the board would fly with that pilot to give some tips and feedback. It would be over a few days so they had time off to chat over dinner or a drink in a more relaxed atmosphere. On one occasion, we flew down to Alicante, a frequent destination, shot the approach and landed. The first officer had flown the magenta line all the way down and did a good job. On the way back, I said: 'We landed on the easterly runway.' Yes. 'You briefed me our alternate was Valencia, so if we couldn't land at all because the aircraft in front blocked the runway, we would divert to Valencia.' Yes. 'Have you been to Valencia?' No. 'Have you looked at the letdown plates for Valencia?' No. 'Let's get them out and have a look.' The runway at Valencia, just up the coast, is in a similar direction as Alicante. In those days, at Valencia there was an ILS on the westerly runway but a VOR on the easterly. Because of the non-precision approach and high ground nearby, the MDA (minimum descent altitude) was high, at 1,400 feet. The charts said you need 1,500 metres in-flight visibility. However, because of the MDA, you would actually need six or seven miles in-flight visibility. The legal chart minima was irrelevant and was not the whole story.

This senior experienced first officer had not considered this. He obviously had not looked at his alternate charts, and, more importantly, analysed what was required to get in. If, on short final approach, we had to divert, we'd have headed off to Valencia and he would suddenly have thought 'Shit! It's a VOR approach, not an ILS. We didn't have enough fuel to make Ibiza, we were committed to Valencia. If the visibility was a few miles, we would get down to MDA but not see the runway. Then what? Now short of fuel, the best option would be to fly the ILS and land with whatever tailwind there was, legal or not.

The learning point of all this was that this first officer was quite comfortable in his cosy blanket. He knew Alicante and had been there many times but had not really considered the limitations of his number one alternate Valencia when landing in that direction. What is more disturbing is that during his ten years in the co-pilot seat, no captain had briefed him on the problem with Valencia. Worse still, no training captain had briefed him. When I told this tale in a classroom of training captains, there was a hush. We were just flying the magenta line to the blue circle and thinking we were doing a great job. We were becoming remote from the jet. When I joined Britannia, monitoring was not front and centre, as it is today. It was all about knowing your individual role on the flight deck. Monitoring was part of it but the roles of the two pilots were separate. This was how things were in the old world where, as pilot flying, you would fly the aircraft while the co-pilot carried out his own drills. The emphasis was not on monitoring each other because, in those days, you were only concerned with doing your role correctly. The words used then were 'pilot flying' and 'pilot not flying' and that's changed to 'pilot flying' and 'pilot monitoring'. As procedures changed, monitoring was built into the operation. For example, you've just turned the engine anti-ice off instead of the centre tank fuel switches. They're both the same design of switches and above one another in the centre overhead panel. Is that what you were supposed to do? So the term became 'pilot monitoring'. In the simulator, the first question from the training captain might be: 'Who is the primary pilot monitor?' The correct answer is: 'Everyone!' But the answer is often 'the pilot monitoring'. Of course, to monitor someone, you need to be actively involved in the process. If you don't know exactly what the other pilot should be doing or trying to do, you can't monitor effectively.

There are two ways of running a flight deck checklist. One is the 'read and do', which tended to be the old British way. For example, the after-take-off procedure might include turning off the nose wheel light and the 'pilot not-flying' would read that on the checklist and then 'do it'. Boeing, on the other hand, used 'scan flow'. For the pre-start procedure, which is always completed by the pilot in the right-hand seat, you would start up in the top left-hand corner of the overhead panel and work your way down in columns, switching on the necessary systems for an engine start. Then work your way along the mode control panel. Once that's complete, the captain calls for the 'before start' checklist to confirm that everything has been done. If there is a reason you have not done something, you would hold the checklist at that point, do what was required and continue the checklist from the omission. If there is ever any confusion, you would start the whole checklist again. That's the philosophy Boeing and Airbus use for flight deck procedures.

Even in the air, when you thought about doing the descent checks, you would first work out the landing weight and decide what flap you were going to use, and then set the speed bug to the appropriate speed. Then you'd brief the other pilot how you were expecting to fly the approach and what charts you will use, set the MDA (minimum descent altitude) altimeter bugs, set the touchdown runway elevation on the pressurisation panel, set the navigational aids for the approach and the destination QNH on the standby altimeter. When all that was done, you'd call for the descent checklist, which the pilot monitoring would read.

When deciding the landing weight at that stage, the only variable is the fuel remaining on landing. The gauges are in weight in kilos, but on the Jumbo you couldn't actually see that. The Jumbo fuel gauges were over on the engineer's panel, which as a first officer you couldn't see because the panel was behind you. The captain could look over and see the gauges. In the Jumbo one time, we were being radar vectored all over the place on the way into Miami and I was becoming concerned about the fuel remaining. No one else seemed bothered and fuel remaining wasn't mentioned when, again, radar gave us another heading away from Miami. No one was making a decision until eventually I had to say something. We were still high, we were nowhere near final approach, and I asked the flight engineer how much fuel we had. The engineer said 'Enough' and I looked forward, thinking: 'No, no, no!' I turned round and asked him to read me each gauge. I said: 'We need to do something – now!' The captain, to his credit, thought that was enough information to take a decision. I said to Miami ATC: 'We need to make our approach now or we will declare a fuel emergency.' We were still off the coast at altitude and we had just 10 tonnes of fuel left - and you want to land with 10 tonnes to be comfortable. ATC obliged, the captain flew a lovely tight circuit and it was fine. When I looked back on it, I could see we were in La La Land, but at the time I had to be careful because I was not the captain. It was a typical situation with two old boys working it their way. I was the only one thinking about the fuel burn.

You can actually fly without serviceable fuel gauges but the engineers have got to get steps, climb up under the wings where the wing fuel tanks are, and pull down a series of calibrated rods. When fuel starts flowing from the bottom of the rod, that indicates the fuel in the tank. Not a nice job and it's the captain's decision whether to accept the aircraft or not, even though it is legal to go.

In 1990, I was flying into Girona and on the approach I was looking at this huge thunderstorm ahead on our final approach. Girona is an awkward place anyway and I said to the first officer: 'I'm not very happy. What do you think?' He wasn't sure either, so we diverted into Barcelona. One of the management team was also on the flight deck, but not a pilot manager - something to do with fuel management. We diverted in and were on the ground for a couple of hours. The passengers were bussed over from Girona and, after a delay of a couple of hours, we flew back to the UK. Other aircraft at the time were landing in Girona, and afterwards I thought: 'Was I too timid?' At some stage you have got to know where your comfort zone is and I considered the chances of us being hit by lightning, causing damage to the aircraft, were just too high and landing conditions were too demanding. All pilots have their own comfort zone and this is something you just can't teach. You've got to be empowered to take decisions like that and you've got to have a sort of worldliness to make the correct call. My decision to divert was never questioned.

On March 4th 1990, as a captain, I was told to take a B737 empty to Tel Aviv because they required some maintenance to this particular aircraft. The B737-200s were getting quite old and although Britannia did the vast majority of the maintenance, lighter maintenance would sometimes be contracted out. Britannia had rebuilt an entire B737-200 at Luton during heavy maintenance and Boeing was quite happy for the company to have a sign written on the side of the fuselage 'Built in Britain'. No one had ever done this before, but we were having to outsource some of our maintenance to a company in Tel Aviv for a while. We landed there and I thought: 'The last time I was here I was told never to come back!' Getting into Israel is always a big deal at the best of times because they have always been very security-minded. Fortunately, there wasn't anything in my passport to record my past run-in with the authorities - but my memories of that encounter kept me on my toes.

From Tel Aviv, we caught a scheduled flight back and that was the end of that little exercise. The next flight I did, on March 7th, turned out to be far more interesting because it was five sectors. We flew over to Paris Orly the night before. Then we did Orly to Marseille, to Orly, to Perpignan, to Orly, and finally to Manchester for an internal French airline which operated Airbuses. They had been on strike, though not the pilots, because our union would have got involved and had something to say about it. Britannia had done similar work before when we were based in Sardinia for a while and also there was a contract to base one or two aircraft in Yemen.

These contracts brought in needed revenue in the winter months when the airline was losing money. The company could only roster you to go for up to 28 days, so usually they would ask for volunteers and there would be provision for the family to go with them. The day flying backwards and forwards in France was great fun. Alan, my first officer, was ex-military and an experienced pilot. We arrived at Orly and we were in Golf Alpha Victor Romeo November, which was one of the old

B737s with the smaller engines. We were full on all flights - so much so that on the Orly-Marseille leg they asked if they could put a passenger on the flight deck, which we would never normally have done. I asked the airline's agent about us keeping time. In other words: 'Do you want me to save fuel?' The answer was: 'No, no, no. We want you to keep to time.' We were flying an Airbus route and the Airbus flew considerably faster than the old B737-200s. We arrived back into Orly four sectors later, before returning to Manchester five minutes early.
There was one occasion when one of our B737-200s was operating out of a southern Arabian country. On take-off, the crew thought they had an engine failure because the aircraft wasn't climbing at all well. Examining the characteristics of the aircraft and looking closely at the engine instruments, they realised that both engines were fine but the aircraft didn't want to fly. They struggled around a visual circuit and landed back on the runway they had just departed from. Opening the holds, they discovered pallets of tonnes of gold bullion. The aircraft was massively overweight, but the locals hadn't told them for reasons of security. It was their big secret.
It was around this time that I visited Doc Roscoe, the company doctor who looked after the pilots and had a surgery in the main Britannia building at Luton. He had completed some valuable work on understanding levels of radiation and he also did work to monitor and understand pilots' heart rates on take-off and landing. He had advised me to use sleeping tablets when I was having problems getting to sleep in the daytime. He suggested Tamazepam sleeping tablets designed for the military. If you take a Tamazepam and relax, you could usually expect to get four hours' sleep and rested for a night flight. Unfortunately, they became famous as a drug of choice for certain people and were called 'jellies' because they were little jelly capsules. These days, they are in tablet form – no doubt harder to inject.
When I told my own doctor in Olney what the company doctor advised, he said the remedy was absolutely fine. I've always had them with me ever since. I found out later in my career, when I visited NASA to take the sleep deprivation course in Paris, that two body clocks control sleep patterns. One is the circadian rhythm and in that you are switching off as you are gradually getting through the evening, getting ready for sleep. Left to that alone, we would want to be in bed by 7pm, so the brain sends a wake-up signal to counteract the circadian rhythm. That switches off quite abruptly late evening so we hopefully get off to sleep by 11ish. Therefore, when pilots think 'I've got to get up at 4 in the morning, so I'll go to bed at 8', they just can't sleep. The Tamazepam

puts you to sleep in the hope that four hours later you will be into the circadian rhythm and sleep naturally. I learned the hard way not to take them at 7pm or earlier, as I have woken at 11pm thinking: 'Damn, I am awake now and my brain is buzzing away, worrying about trying to get to sleep for my 4am alarm!' Without them, I would go to bed at 9pm and lie there trying to sleep because I knew I would be getting up at 3am.

When Doc Roscoe retired, we had to find our own doctors. He told me: 'There's a chap in Bedford who works for the RAE (Royal Aircraft Establishment) at the former RAF airfield, now RAE Bedford Thurleigh, where they have the huge wind tunnel.' When I first went there, it was still run by the RAE but it has since been sold and built on. I arrived at the front gate and had to go to the guardroom. It was all very formal. I found the medical block and was met by a nurse, who gave me a form to fill in and various papers to sign. I was due for my hearing test and was ushered into a fancy cubicle and given headphones. After seeing a nurse, I saw a doctor who made a comment about my weight. This was in 1990 and I maybe wasn't the slimmest I have ever been, but I wasn't big because I had been doing a lot of running in the late 80s. In fact, I had run until my knees got sore and was told to stop. To my dismay, he pointed to an adjoining room and commanded: 'Right, into this torture chamber and strip off to your underpants!' Obeying meekly, I went in and took my clothes off. I was standing in this bleak little room nearly naked when the doctor walked in and burst out laughing – not just laughing, howling with laughter and pointing at me. It was only then that I looked down and noticed I was wearing my psychedelic underpants that I'd got in Paris. All I could do was cringe with embarrassment and mutter: 'Oh yeah, I got these in, er, France…' It turned out my bright green underpants were not even the big joke. He howled: 'I'm not laughing at your underpants – you are a fat bastard!' I nearly cried there and then.

Despite this crippling humiliation, I continued going to see him for the next 23 years. Even in his retirement, I arrived at his house in Bedford for a medical which usually took the whole morning. He checked my weight as well, of course. As soon as I got to the front door, he would always shout out to his wife: 'It's the fat bastard!' I've had to say to him: 'I'm paying a couple of hundred quid for this abuse.' He'd reply: 'But you're still a fat bastard.' Over the years he gave me a really hard time about my weight - when I phoned up, I'd say to his wife: 'I'm not coming unless you're there to protect me!' However, we really did get on famously. In 1990, I had my first line check. We were all line checked every year back then, which meant a training captain would sit in the jump seat and watch you closely. It was usually to Dusseldorf, or one of the other trooping flights, so the trainers weren't sitting for too long on the B737's uncomfortable jump seat. The trooping flights were quite a demanding day out.

That first line check on March 15th, Luton to Dusseldorf and back, went off fine. I was with a second officer although, normally, a new captain would not fly with one of them, but I had obviously done enough that I could. We climbed over the North Sea and, of course, there was very little cruise between Luton and Dusseldorf. The efficient way of doing the flight was to keep climbing and then close the thrust levers and go straight into descent. It is also the most difficult because there is no cruise time to get your head sorted out, plan and brief the descent and approach – and, more importantly, eat breakfast. We were climbing and were indicating about 20 knots fast, and we were going to stop at flight level 270. At about flight level 250, I hadn't said anything to the chap who was flying it, but I said: 'You're a wee bit fast.' He said sorry and pulled back on the control column. The 20 knots of energy was converted into a phenomenal rate of climb and I thought: 'Shit, we're going to bust the flight level.' I had to take control. The training captain was sitting in the back and I could feel the smile on his face.

Under new management

One of the major events of 1990 was the arrival of a new management at Britannia and the end of the era of pilot managing directors. Owner Lord Thomson thought we needed to make more money and the company had become a bit of a pilot flying club, so there was a senior management shake-up. All the staff arrived at Luton Hoo, a stately home near Luton airport, for a series of huge meetings where the message was 'It's the end of the world as you know it!'. The new management were going for big cutbacks and were getting rid of the old B737-200s. They were going to bring in B757s, which would mean fewer aircraft – and fewer pilots. We had all been working extremely hard, with lots of flying hours, and the union had been negotiating to prevent compulsory redundancies. They were talking about something called 'stand down', which was only available to captains. The idea was that you could leave the company with half of a redundancy package, which they would give you straightaway. You would be kept on a list for 10 years, meaning you could come back at any time if there was a job available and you would return with the same seniority. After 10 years, or at any point when you said you would not be coming back, you would be sent the other half of the redundancy. It was a pretty good deal, and it made me think, as I had always fancied flying with a flight engineer.

At that time, Richard Branson's Virgin airline hadn't been operating for long. It started with just one aircraft and in the early days it was a right mess. This old Jumbo, a B747-100, would arrive back into Gatwick three days late from Miami. But, despite all their troubles, they kept going. When the passengers came off a nightmare flight, Richard Branson would be standing at the aircraft door with a bottle of champagne and a free ticket for everyone. By the time these stressed-out passengers got into the terminal, they were the happiest people in the world. They had met a celeb and everything was forgiven. It could not have worked anywhere else. Virgin was looking for pilots as they had built their fleet to six planes, all of them B747 Jumbos. They had one 100 and the rest were 200s. These aircraft were the same size but the early 100s tended to have a bar area and a few seats upstairs. The 200s had bigger engines, and the later 300s had the longer top deck. It still had three flight crew. Then the 400 came out, with the same upstairs but without the spiral staircase. Boeing built it like the B767, with modern technology and a two-crew flight deck, losing the flight engineer.

Along with my pal Stu Moffat and a few others that followed us, I decided to join Virgin Atlantic. Most of the fleet had come from different stables. The Miami B747-100 aircraft, surprisingly, had more than 500 seats because they had taken all the galleys out and they were re-situated in what would have been part of the luggage hold. There were lifts up and down to the galleys. The whole main deck was just seats and a few toilets. Because of the Britannia deal, I had the opportunity to go to Virgin with a sort of free ticket and a bunch of money in my pocket. From my point of view, there was no other airline I could have gone to that would fly out of the UK with a flight engineer. I had decided that I definitely didn't want to move around the world. There were compelling reasons to contemplate a move at that time because there were going to be redundancies. The phones were red hot as the company and BALPA went through almost daily meetings.

At one stage, it all became a bit ridiculous and the company was talking of around 100 pilots going. It was suggested that they wanted to get rid of 100 first officers and demote 35 captains. There had been demotions before, so this was not a completely new concept. In the past, the demoted kept their rank and salary, just like I did years before at Loganair. They felt they had too many captains and, generally, a captain couldn't fly as a co-pilot. The only captains who could fly as co-pilots were training captains, because pilots had to be checked out in both seats in the simulator. In a perfect world, the management would have liked to get rid of 50 captains and 50 co-pilots but they couldn't because the union had in place a 'last in, first out' agreement and would fight tooth and nail to preserve it.

As the company started the costs and staffing review, I had to calculate where I was likely to line up in the pecking order. They would start at the bottom and I had been there eight years, so I considered I was quite safe. On the other hand, I hadn't been a captain for a year at that point.

I was in the demotion zone. It was not a happy time for aircrew and feelings were running high, particularly because management wanted to get rid of pilots almost as they wished and not in accordance with the long-held agreement of seniority. I was a strong BALPA member who believed in seniority. In my union days on the company council, I even fought against my own interests sometimes but I believed more in the principle I was fighting for. The company let it be known that they wanted to get rid of people by rank and possibly by base.

The summer went on with me pondering: 'Should we leave?' My friend Stuart and I were quite excited about the prospect of heading off on a new adventure. We had never flown long haul or the B747, but we did know deep down that Britannia was a far more solid airline than Virgin, and likely to see us out career-wise. But if we did leave, we would have a return ticket. As I considered my future, I had to accept that I didn't really know what long haul was. My entire career in Britannia had involved going to work and then, five days later, coming home. My assessment was that it wouldn't be much different. How wrong I was. I had never even experienced the dreaded jetlag up to that point.

Ruth was quite happy letting me have a go at long haul. I decided it was time to take control of my own fate rather than waiting to be told 'you're in Glasgow now' or, worse, 'you're demoted'. Camille was by now a first officer on the Jumbo with BA, flying all over the world, and he used to talk about Singapore and other far-off places. When I did leave Britannia, I didn't leave thinking: 'I'll be back.' I left thinking: 'The next chapter in my career!' They gave us a month's salary for every year we'd worked there. I went in to see the chief pilot, and I ended up getting more like a year and a half's salary. He said that in my position, he would have been off in a flash.

Stu and I went to Gatwick for the Virgin interview to see their chief pilot Tony Ling. All the Virgin pilots had joined with lots of experience on the B747. They were all type-rated and were trying to get a job back in the UK. A lot of them were ex-Cathay Pacific. If you had a career with Cathay Pacific in those days, you left with about a million pounds for your pension pot but you didn't actually have a pension. They also came from BA because their pilots had to retire at 55 but their licences were still valid until 60. Many went off to various airlines to build up their pensions or just try something different. A few British pilots had also come back from Saudi Arabian Airlines. They were all an experienced bunch. We were going to be among the first pilots they brought in who were not B747-rated and therefore needed B747 type rating, because they obviously couldn't get enough type-rated pilots.

There might have been more to it than that. It might have been more about building the Virgin culture on the flight deck. Each airline has a different culture and also, hopefully, a strong safety culture. Captains from various airlines brought in different ways of doing things, while promoting co-pilots from within with experience of the airline's culture was also a key part of the process.

We got the job, even though we told the chief pilot we could return to Britannia. He replied: 'It's our job to keep you.' We began the simulator course in November. Virgin couldn't do its own aircraft technical exams so we had to go to the CAA for that. There were four of us on the course, all from Britannia. I was swotting the whole weekend, but the others were extremely bright individuals and very adept at taking in vast amounts of information. After weeks on the course, we sat our exams and all achieved 97%. I was pleased because it showed again that I could achieve very good exam performance – I just had to put in more time and effort than the bright boys.

The Virgin B747 ground school was more like the basic manufacturer's course, fairly minimalist but enough to get us all through the final exam. Initially, there had been a chance that we would join as direct-entry captains. My thought was: 'It's only a Jumbo, we can do that!' But the first Gulf War kicked off around that time and it doesn't take much of a war for the Americans to stop flying. Virgin, of course, was a schedule carrier and, with the exception of Tokyo, all the destinations were in the US so they were taking a big financial hit at that time. Early on, they told us: 'You're going to have to start as co-pilots.' We thought 'fair enough', but there was a pay cut which we had to accept. We had taken a pot of cash on the way out at Britannia so financially we were fine.

The difference between Britannia and Virgin was apparent almost straight away. Virgin was like a little flying club - it was lovely and laidback. When we were talking about landing performance at Britannia, it was all about wet or dry performance and they would talk about PCN (pavement classification number) values, which gauged whether it was possible to land a particular aircraft on a particular runway. When I began asking questions on these subjects at Virgin, the answer was: 'Oh son, you're all right. If you can get a B707 in, you can get a Jumbo in.' They had all flown B707s but it was meaningless to me. It was all a bit vague and left me feeling vulnerable.

We went to do the simulator training in Miami, using Virgin trainers but on the PanAm training department simulator - an exciting prospect. The difference with Britannia was again apparent. Every question at

Britannia seemed to provoke the reply: 'That's sensitive information – we don't want other people hearing that.' Virgin wasn't like that. We were in the classroom during the ground school and the engineer was talking about one of the aircraft systems when the operations director popped his head in and said: 'You've got your courses in November in Miami. Are you taking your wives or girlfriends?' Someone said: 'I thought you've got to be in the airline six months before you can get a cheap ticket.' He replied: 'Don't worry about that. Have you looked to see what the loads are like for the flight you are due to go out on? Do you know how to get on to the computer system? No? Okay, follow me.' We all left the engineer to talk to himself and followed the ops director into his office, where he said: 'Gather around, the door is always open. Do you know the code to get into the system? No? Well here is mine.' He gave us each a code - and added: 'There's loads of room on the Miami flight.' Ruth could come with me and her parents would look after the girls. Ruth, Stu and I arrived at check-in at Gatwick for our Virgin flight to Florida. Stu and I had confirmed tickets and Ruth a stand-by ticket. They immediately upgraded the three of us to Upper Class and said we could use the lounge. This wouldn't have happened at Britannia – they didn't even have a lounge!

The Jumbos were old but the outsides looked smart and the interiors were immaculate. Richard Branson had taken aircraft which many regarded as 'old duffers' and had them looking really good. The crew looked smart as well. We were sitting at the front of the aircraft, downstairs, there was a bar with bar stools, and the crew came around with 'the wine basket'. One of the girls asked if we wanted to go up to the flight deck, and Stu and I said almost in unison: 'No fear, this is marvellous. We will be spending years up there!'

Down to business

Before we started our first day in the simulator, we had a briefing in the training captain's hotel room. He explained that we would do some three-engine work and finish with an NPA (non-precision approach). When we asked 'How do we fly the profiles?', he said: 'Just follow the Boeing profiles.' On hearing that, Stu and I went to our rooms in a blind panic because in Britannia we used Britannia profiles, which were far more detailed than the basic Boeing charts. There were details at every point of the profile for a given weight, speed, power setting, aircraft attitude, plus all of the required calls and responses for the given profile. For example, it might have said, as we crossed the beacon outbound: 'You should be Flap 5, speed Vref flap 30 plus 40 knots with 5 deg nose up and engine power at 62% N1. And if you do that, the aircraft will fly level.' Page by page, we would have the information to fly each profile, eg. an ILS, an NPA or a visual circuit.

It was all laid out in a diagram for us to learn because a jet is actually flown by the numbers. Suddenly, at Virgin we had no numbers and we couldn't learn it because they didn't have any numbers to give us. We were going to be made to look stupid.

We thought we'd just use the B767 profiles we still had from Britannia. We returned to one of our rooms and tried to compare the Boeing 747-200 profiles with the Britannia Boeing 767-200 profiles and cobble something together. The PanAm simulator was a different configuration to the Virgin aircraft, which were all different anyway. We were used to lots of detail, not comments like: 'It will be okay, don't worry.' It felt a bit of a shambles. We arrived with our B767 stuff under our arms – and at the end of day one, the training captain took us aside and asked: 'How did you do all that?' We confessed that we just did it like a B767, and he replied: 'Boy, that was good! Can we borrow these things?' We had 10 lessons in the simulator over the next few weeks and by the third lesson they were saying: 'We're wasting our time here. You guys can do this. Why don't we go down to Key West for a day or two?' Good plan!

The B747 flew very like the B767, although it was not as powerful, but the general handling qualities of the two Boeings were very similar. Sadly, our engineer wasn't doing quite so well. We had flown the Jumbo like a B767, and it handled like a B767, but when it came to sheer power, it was gutless by comparison. I noticed early on that whoever taught us had his own way of doing so, which meant there was little standardisation. Trainers taught you the way they had been taught in their previous airlines and often used their previous standard operating procedures. For example, they would fail an outer engine, which would be one of the more critical engines to lose at V1. This was due to the greater swing because of the larger moment arm of the outer engines. One trainer would say: 'Just fly it like a two-engine aeroplane.' As you increased power on the remaining three operating engines, you needed to apply rudder to keep the aircraft straight. Then someone else would say: 'Don't do that. If the number one engine has failed and is shut down, leave a little power on the number four and just fly using the inboard engines numbers two and three, because then you have no asymmetric yaw.'

Stu and I would brief each other and agree 'let's do it this way', and it worked out fine. We were lucky that we had been well trained and disciplined in the early parts of our careers. We had already put the work in, and we made it work at Virgin and did well. It was noted by the Virgin trainers, but to us it was no big surprise.

Boeing produced basic flight patterns but they were basic. Boeing are not trainers, they build aircraft. They train pilots enough to sell the aircraft, but even now if you saw a profile of a B767 from Boeing, and then looked at an airline's profile of the same aeroplane, there would be much more information in the latter. All in all, we were under no great pressure during those early weeks - in fact, looking back, I think our biggest challenge was getting through the enormous American breakfast every morning.

We would go to the simulator at Miami Airport but always seemed to end up getting the graveyard shift because their own pilots were getting the best slots. We usually visited the Pan American training crew rest area, and all around the room they had hundreds of black and white pictures, starting with the flying boats PanAm was famous for. The history of the airline was fascinating. We learned later how bad things were getting at PanAm. We had finished our course but a few courses after us there was a problem. The flight crew were training in the simulator and suddenly all the power went off. The whole sim fell down from its jacks and when the crew emerged, there was only emergency lighting because the place was in darkness. Rumour had it that PanAm had gone bust, so the electricity supply company pulled the plug.

One thing we quickly noted at Virgin was that we weren't 'pilots' any more, we were 'flight crew'. If you talked about 'pilots', the engineer could get a bit upset and say: 'What am I, chopped meat?' I arrived back in the UK in December to do my line training, which, with Virgin being all long haul, took a while because you couldn't bash in four sectors a day. First, I went on a 'famil' flight to New York - to Newark, where I'd never been before. All those years before, I had flown over with Jock Adams and I'd had the engine failure in the DC-3 above New York, but I never had to land there. In those days, Virgin flew to Newark, JFK, Miami, Orlando, LAX, Boston and Narita (Tokyo). I also had to go base training and fly the Jumbo empty because the simulator was old and back then there wasn't zero flight time training in a simulator. You had to fly it for real. They took an empty aeroplane to Shannon and I flew it out with the training captain.

The flight deck of a B747 is long and narrow, and high up off the ground. You're up in the bubble and when you're in your seat, your head is almost touching the side wall panel and side window. The flight deck opens out behind you. The engineer sits behind the first officer, facing to the side, but if I am sitting forward, he is over my left shoulder so I can't actually look to my right and see anything. I have to look left and if I do, I can't see the flight engineer's instruments. The flight deck was long because when it was first built, the B747 would have carried a navigator as well as the two pilots and the flight engineer. So there was lots of room. The flight engineer's seat was on rails, so for take-off and landing the seat could swing to face forwards and move behind the central control panel, between the two pilots. There was no separate set of thrust levers at the engineer's station, as in some of the older aeroplanes, and there was no real auto throttle on the aeroplane. It was fitted but it was very basic and our aircraft were old. The result was that they weren't very reliable, so the engineer worked the thrust levers until final approach, when I would take over. Beside the engineer's panel and seat there was an escape door. There was also an escape hatch in the ceiling, which was counter-balanced. You would put on a rope around you and under your arms and leap off the top, and it would let you down at a certain rate. Because of the height of a Jumbo, I always thought: 'If I have to jump, the engineer's going out first because there would be something soft to land on.' There was one nice touch: a Perspex dome on the roof in the flight deck for taking star shots. In the early days, navigators would take star shots to navigate by, before ground beacons were established in remoter parts of the world. This was the post-war period of early jet aircraft. A lot of long-haul flights were at night so they could take star shots to navigate by. The flight deck had a bunk on the left, going back behind the upstairs passengers' bar. On the right, as you entered, some aircraft had a toilet for the flight crew. I never felt it was a particularly comfortable flight deck. Although it was long, it didn't feel spacious, whereas the B767, at the front of a big wide aircraft, was much more so. I had a good couple of feet between me and the side window.

At Shannon, we did circuits, flying the aircraft out with the trainer. I thought: 'I didn't look at this enormous great thing, did I? So I'm not intimidated, am I?' I needn't have worried. It flew just like a B767. However, for the flare (lifting the nose before landing) you were high up because, first, it was a big aircraft and, secondly, you were high up anyway. When the RADALT said 50, 40, 30, at 30 feet you flared. You didn't look out of the window and think: 'Am I still a bit high?' If you did, you would bury the jet into the runway. In crosswinds, if you had a bit of wing down, you had to watch the outer engine pod, which could be a bit close to the ground. But, overall, it wasn't a demanding aircraft to fly.

The next flight in line training, we flew out to JFK – I had never been there either – and I was thinking: 'I'm actually glad I'm not a captain. This is such a different operation to flying around Europe.'

The workload was low because there were three of us. The two pilots could concentrate on flying, and if we needed weather information or different charts, there was another pair of hands in the flight deck. Engineers were very much part of the crew. They were a balance of old guys from BA and younger ex-RAF engineers. They read and managed the checklists once the aircraft moved. They were also listening on the radio and held RT licences. It was a proper three-man crew, a bit like flying with two officers and a sergeant. They weren't trying to impress the captain and hoping to move into the left-hand seat. They were generally happy to speak up if they thought something wasn't right. Some other airlines treated their engineers as just engineers. They didn't do any of the checklists or operate the radio. On one occasion, on a southern hemisphere airliner, an engineer was sitting over the Pacific and the two pilots were missing a call from ATC because they were chatting. The engineer said 'Perth is calling you' and rather than saying 'thanks', the captain gave him a bollocking for listening in on the radio. The engineer we went to Miami with actually failed his course and didn't join us.

Virgin crews were mostly a lovely bunch. They were big crews, with maybe 15 cabin crew and sometimes five on the flight deck. That meant you seldom all went out together as a crew. The amount of crew we had varied because the cabin configuration changed, depending on the destination. Some of the aircraft had bigger upgraded engines and they were used on the Narita and LA routes. The Miami route had just the basic old B747-100, which had no posh seats. They kept us on the shorter routes to try to get the sectors in during the line training.

It was quite challenging coming into LA and the line training captains were not terribly proactive. They just let me get on with it and then gave me some tips, rather than saying: 'You haven't been to LA before. This is how you do it...' Maybe that was because they weren't used to teaching new pilots with no B747 experience who had not flown in the US before. If you had been flying B747s before joining Virgin, you had no doubt been to LA a dozen times before. Once, in my first year as a first officer, I was asked by ATC to 'side-step' on to the parallel runway. I was only three or four miles out. I tried but failed to get stable on the final approach to the new runway and had to go around. ATC were trying to be helpful by giving us a much shorter taxi route to our gate. My captain phoned the tower when we got back and had a blazing row: 'What do you think we're flying, a f****** Cherokee!' It was all a big learning curve. By February 18th 1991, I had done my final line check, back from Los Angeles in 11 hours - and then I was let loose on the world.

Like a Virgin

One of my first flights after my final line check was to Narita, which we did in 13 hours 15 minutes. The aircraft was Golf Tango Kilo Yankee Oscar and had the big 'Q' engines for the route to Japan. Virgin had taken delivery of two more aircraft, both from the same stable and both with the 'Q' engines, with more grunt and able to carry more weight. The fleet was now up to eight. As we taxied on to our stand at Narita, I saw all these Japanese ground engineers in a line on the tarmac, bowing as we came to a halt. We had to shut down the engines and get towed on to the stand because the turn was too tight to taxi under our own power. They were so efficient that by the time we cut the engines and were running through the checks, they had informed us that the tug and tow bar were attached and we could release the brakes to be towed the rest of the way. The captain had been a very senior man with BA, so he used to go to Japan a lot. He had even been given the honorary title of 'Shogun', so popular was he, so when we arrived at the hotel there was a lot of bowing going on.

Because I was on line, I did three long-haul trips a month. The airline would try to give me an LA and a Narita – quite challenging because of the time changes – then a short one to Boston, New York or Miami. Narita used to come up once a month and I soon discovered that the majority of the cabin crew didn't like it because everything was so different. They brought out loads of pot noodles and microwaved them in the crew lounge in our hotel. The only local product they consumed was the beer. It was not uncommon to stand looking at a shop which looked like it hadn't changed in the past few hundred years. The sights and smells were all strange: I would look into a shop that obviously sold edible stuff - but edible by who or what? I did not recognise one thing.

The Narita flights left Heathrow about lunchtime, we'd fly all night, and when we arrived, there was a colossal time change, with the clocks going forwards by nine hours. The hotel provided a large room for the crew to relax in. Richard Branson said: 'If you want to take drink off the aeroplane, that's fine. Take what you drink.' There were lots of open bottles left over from the flight, so if you wanted whisky or anything, you carried a hip flask and the crew would fill it up for you. Branson wasn't so happy with drinks being taken off an aeroplane coming back into the UK. He was insistent on that point and said: 'I'm not there to fill up your drinks cabinets at home.' We arrived at the hotel in the afternoon their time, went to the crew room for a beer, and then went to bed. By then, you would have been up for 20 hours or more. The plan was to get a few hours' sleep, then get up at 6pm and catch the hotel bus down town.

If you slept on and didn't go into town, you would wake up at about 11pm and be wide awake all night. We would go to the English bar where you'd have a couple of beers and something to eat. Then we'd go to the karaoke bar, which was completely new to me at the time. The crew would sing their hearts out while consuming a fair amount of the local Sapporo beer. Again, everything was strange and different, and mostly wonderful. In the karaoke bar, the ladies' toilet was accessed by walking through the gents. So we'd be having a wee and our girls would pass and give us a shove on their way through.

The Truck was famous around the world with aircrew. It was the back of an articulated lorry which had been converted into a den of iniquity for aircrew on stopovers. It was full of debauched aircrew who would bang on all night - an engineer once let off mini-rockets out of someone's back pocket. You'd leave there pissed and occasionally a crew member would be lost on the way back to the hotel and wake up at the bottom of a storm drain. I didn't have the stamina to go out clubbing all night without feeling dreadful the next day. One beer in The Truck and it was all over for me – I had to go to sleep. It was always very noticeable when our crew arrived – five from the flight deck and 15 girls all dressed in the Virgin bright red (and, in the early days, all blonde). The Virgin TV ad of the time really did portray what it was like. We would walk through terminals or hotel lobbies and everyone stopped and stared.

When I was training in Miami, some of the training captains were very senior management within the company and also flew the line. We used to go to the Irish bar in Miami and drink there with these senior captains who were friendly and very open with company information. They would be telling us future plans about how we were going to start a new route and screw British Airways by doing this and that. We would be listening to all this, thinking: 'They're all really pissed, what's going on? Have they forgotten we're at the table?' I quickly realised that the company was open with its plans and what was going on in the back office. It was, in a way, very empowering, but really, at the end of it all, you actually learned nothing new – everything is so fluid in aviation that the plan was changing day by day.

Branson himself was very popular with the cabin crew. When we flew with him, he would always go out with the trolleys and serve the passengers, who loved it. He'd come into the flight deck and say hello, and then leave us. His friendly nature didn't always go down well with all the flight crew. It is said that in the early days Branson was rebuffed by a captain he had embraced and that subsequently he spent less time with the flight crew and more with the cabin crew. This applied both on board the aircraft and to those times when he would take the whole crew out in the evening. He needed his flight crew, of course. They were part of his train set, but he didn't spend a lot of time in the flight deck.

It was different with Freddie Laker, who frequently flew with us. At that time, his airline had already gone bust and quite a few flight engineers were ex-Laker and ex-DC-10. He was running a new company in Miami, flying B727s around the Caribbean, and would often travel out with us to Miami. The difference was that he'd spend the whole flight in the flight deck. Richard was there to put the gloss on the girls and the aeroplanes. After the Laker collapse, people weren't paid but that didn't seem to affect his reputation among the flight engineers now flying for Virgin. Freddie would walk in and say: 'Do you mind if I come in?' He would recognise the flight engineer and say: 'Hello, John.' And although John had lost money and his job because of Laker, he would always be delighted to see Freddie and they would chat away.

Laker and Branson showed how very different the experience of two prominent businessmen in the 1990s could be. Once a year, Richard would have a party for his Virgin staff in the grounds of his house in Kidlington, Oxfordshire. It used to be held in the house but as the airline grew, it was moved into the garden. He'd have a helicopter brought in to give rides for the kids. Three of my girls went up to Richard during one party and thanked him for being invited. He was carrying his son on his shoulders at the time and he said to them: 'Thank you for coming.' He was charming.

The crew stayed in some tremendous hotels, including the Peabody in Orlando, which is famous for its ducks. They are brought down in the lift and they all waddle to the big fountain in the lobby. For some reason, the crew nicked the ducks and ransomed them one day. Needless to say, we were kicked out not long afterwards.

In Britannia, you couldn't take your wife on the flight deck in those days, although later this was changed, and you couldn't take your children with you on a trip, even in the cabin, until they were 17. At Virgin, it was 14, and I said to someone: 'I'd like to take Inga, my oldest daughter, but she's only 12.' The reply was: 'That's close enough.' So I could take them one at a time when they were old enough. They'd get their own posh seat and sometimes sit in the flight deck.

By the spring of 1992 I had to make a decision. Do I stay and get my command with Virgin or do I go back to Britannia? Britannia had phoned a few times to ask if I wanted my job back because they couldn't promote anyone until they had offered it to me. My friend Stu had already decided to go back. He flew the line for only a couple of months with Virgin and then went back because they were offering him his command back and his captain's salary, so financially it was worth it. However, there was a bond with Virgin which meant I had to pay for the type rating I had obtained with them unless I stayed for three years.

All my neighbours thought I'd joined a 'proper airline', not a 'charter airline', whereas every Virgin captain I flew with said: 'You don't want to wait around here because you guys can go back. What the hell are you waiting for?' But I was enjoying my time at Virgin and thought: 'I don't want to go back yet!' What started to niggle for the first time was that we had given up a damn good pension. I'd never really thought about the pension before, but the blanket had gone. Now I was in a private scheme, whereas the Britannia fund was one of the best in the UK. A colleague and I went to see chief pilot Tony Ling and he said: 'I can't promote you now but you should be captains by the end of the year.' In the end, we thought we'd had great fun, Ruth and I had seen parts of the world we'd never seen before, but boring old Britannia was where the pay was good and the pension was great. By then, we knew the B737 was going to be replaced, and the B767 was doing long haul. I guess we knew we weren't giving up long haul forever.

I wasn't with Virgin long. I did about a thousand flying hours in less than 18 months on the B747-100 and 200s. I got a call from Tony Ling, who said: 'Someone told me you wrote the ETOPS course for Britannia. Do you still have it?' I said yes, and he replied: 'Would you like to manage a B767 fleet?' That morning Richard Branson had been on the news announcing: 'We're buying B767s.' Tony was already getting calls from people saying: 'I'm a B767 captain.' It turned out that the whole thing was just flying a kite. Richard hadn't been in the press enough that week.

Britannia said we could come back as B737-200 captains – but in Newcastle. Of course, we thought this was a punishment posting. The salary was going up to that of a captain, but the timescale for being in Newcastle was uncertain. No one could say, but I was guessing it would be for a couple of years. Our position now was quite clear. It was about finances and it was about pension (we were becoming old and boring). Looking around Virgin at that time, I realised it was never going to be Britannia, but it was a happy place to work and the crew were proud to be part of the Virgin brand. I, with one of the captains, started a union in Virgin. There was a flight crew association and we had put in a vote of no confidence. BALPA wasn't very popular because the ex-Novair air crews felt let down when their airline went bust in 1990, and thought the union hadn't done enough for them. However, we went to BALPA and asked for help. We got it up and running and we used to have meetings with the company. Historically in aviation, being in BALPA was not a bad thing, especially when pilots ran the place. More recently, Ryanair's Michael O'Leary fought tooth and nail to keep BALPA out and Loganair too tried for ages. They can't stop pilots being in BALPA but they can stop them having representation within the airline. My co-conspirator and I talked through our plan to push for a vote of no confidence in the Virgin flight deck association. I had no doubts that we needed a new association – the flight deck crew needed proper professional representation.

I went into the operation director's office after one of our meetings. He was looking at a pile of pilot applications and said anyone was welcome to go through them. If we knew someone, we could take out the application and put a sticker on it to indicate this and if we thought he or she would be an asset. Not a bad starting position. He obviously felt intimidated by this large pile of applications so he picked up about half and dropped them on the floor with the comment: 'I don't like unlucky people!' The new union was started and accepted, and BALPA gave us great support. No union is perfect but unless you work for a benevolent company, and I have yet to find an airline that fits that bill, it's all we have got. As in the Novair case, the union would not always save their bacon but usually it made a difference. The union's effectiveness always reminded me of my horror experience on top of that Scottish hill in the heather with Loganair. At that moment, a time of real difficulty, I realised I had no one to talk to.

Exhaustion

When we were first interviewed at Virgin, the plan was for us to get direct-entry commands. It wasn't something I was desperate to achieve immediately and after a few Atlantic crossings, I was quite glad to be in the right-hand seat learning this new operation. It was soon clear that the two airlines were like chalk and cheese. True, Virgin was fun but the pay wasn't anything special and the pension was not great. If they had shut down the following year, no one would have been too surprised. It certainly hadn't yet become the big concern it is today. Unsurprisingly, most of the pilots who left Britannia went back. When we started thinking about our return, we knew we would not be in Newcastle forever. Some of the Britannia pilots moving to Newcastle had been newly promoted and were told that was where the command was. There were six or eight of us in all being sent to be based in Newcastle - part of the push to get people out of Luton and into the less popular bases. Some of them had decided to move their families but I remembered what a flight engineer in Virgin had said. He had worked for 10 airlines all over the world, but had always kept the family in one place. That seemed a good idea. I took digs and commuted until either I got moved south to a closer base or I couldn't hack the commute any more. Then we would all move to Newcastle and hope the family would settle.

The hardest part in Newcastle was the first year because there were a lot of standby days. In the summer, we were busy so I'd work for four or five days, come home for a few days and then drive back up, but in the winter I might only do a couple of flights a week. I had to stay there because there was a flight, then two days standby, then another flight, and then another couple of standbys. So I could be in Newcastle for seven days just to do two flights. After the first winter, BALPA sorted out the roster so the company didn't put us on meaningless standbys. They also grouped our flights together so we could stay at home for longer periods, which made a huge difference to our lifestyle.

We had two variants of the B737-200: the dash 9 and the dash 15. The dash 9s had smaller engines and we struggled to make the longer flights without stopping en route for fuel. The company would try to send the dash 15s with the bigger engines to Newcastle to cut out the tech stops. In the early 1990s, TVs were fitted in the ceiling down the centre aisle to provide in-flight entertainment, which made the aircraft even heavier and we were unable to lift the required amount of fuel off the runway at Newcastle when operating the longer routes. Flights were always full and the passengers from Newcastle were delightful. The cabin crew always said that the further north the departure airport, the nicer the passengers. That made Gatwick passengers the worst. On a ski flight out of Gatwick, a cabin crew member came into the flight deck fizzing. She had been serving drinks to a family and the young child said: 'Coke!' She said 'Don't you say please?' - and the mother said: 'We don't say please to staff.'

Operationally, we were really pushing the aircraft in terms of range. Unlike a scheduled carrier, where you could dump people, we couldn't. They had paid for their holiday and they expected to get on board and travel. We'd go for a there-and-back flight and not come back - and it happened every week. I was due to operate a flight back to Newcastle from Izmir in Turkey, but I couldn't lift the required fuel because the aircraft was too heavy and I didn't have the take-off performance to do the flight in one hop. I had to stop in Amsterdam to pick up more fuel. We arrived back in Newcastle hours late, which meant the next flight left late too. This was happening frequently. Flying a busy summer roster is demanding enough, but when there is disruption to the schedule and you are extending your flying duty hours on most of the long flights, fatigue starts to creep in.

Next day, I was meant to fly from Newcastle to Corfu and back, but the plan was a non-starter. We couldn't lift the required fuel out of Newcastle to make it direct, so we had to plan a tech stop outbound. The runway at Newcastle isn't short but it isn't long either. The amount of weight you can lift off a runway is dependent on runway length, slope, air temperature, the air pressure, and if the runway is dry, wet or contaminated. This is exactly why the B757s were arriving, because they have much better performance than the old B737s, but at the time we were still in the old B737-200s. We had to stop southbound in Amsterdam to pick up fuel, which added an hour, by the time you have decelerated, taxied, refuelled with the passengers on board – usually with fire trucks in attendance – taxied out and accelerated. On top of that, each extra sector reduces the pilot's flying duty period by 45 minutes. We made Corfu but again we couldn't lift the required fuel to make it back to Newcastle. The day was falling apart.

I went to our handling agent's office, because we didn't have company mobile phones in those days, and called our operations department in Luton to advise them I could only get the jet as far as Geneva and then I needed to get off. The plan was for me to get there and they would put a new crew in a private jet and position them out to Geneva. Then we would get a lift back to the UK on the executive jet. The chaos to the flying programme was obvious. This was becoming the story of that summer. We would launch from Newcastle, sometimes not knowing where we would stop. If we weren't too short, we would have a go and stop somewhere when it became clear we were not going to make our destination with the required reserves of fuel. We failed to get home quite a few times and the workload was huge and incredibly tiring. You would leave Newcastle thinking: 'This isn't going to work today. I wonder where we will end up.' 'Tech Stop Birrell' became my nickname for a while.

We were opening up Turkey around that time and foremost in my mind was usually where can I get in and out quickly? Sometimes I was just unlucky and I was assigned flights to places like Larnaca, which involved not just tech stops but night stops because we would run out of flying hours. On one occasion, going to Izmir, the B737's maximum fuel load was 15.5 tonnes and I had picked up a flight plan that required 17 tonnes. I simply couldn't get all the fuel on. There was an ATC industrial dispute that year and we were being sent by longer routes over different countries to cut down air traffic delays. We headed over Copenhagen because they couldn't give me a more direct route.

We were never going to make Izmir so we stopped in Bulgaria. Operations called me on the HF radio to say we'd get in and out okay. Wrong! It turned out to be a nightmare. They wanted cash for the fuel and I had to find a bank in the terminal to get US dollars. I should have been in and out in 40 minutes but we ended up there for hours. By the time we got airborne, we had run out of hours. I went Newcastle-Sofia-Izmir and the company told us to get the aircraft across to Rhodes so they could stick pilots on another flight who would bring it back. We got airborne from Izmir for the short flight to Rhodes and were both really knackered. There were no flights from Rhodes for a few days. The company said to send the cabin crew by ferry to Kos, where there was a Britannia flight with four spare seats to somewhere in the UK. There was no way I was putting four girls in uniform on a ferry to Kos, on their own with no company credit card in case it all went wrong. I told the company they'd missed the ferry, and they said: 'Okay, stay there. There's a flight in two days.' Eventually we flew back with Olympic Airways via Athens. The bottom line was they were introducing flights to further destinations and the B757 was needed but hadn't yet arrived, so they were making do.

One fact is certain in aviation: as long as aircraft are getting airborne, there will be a jet where it shouldn't be. This causes the company to haemorrhage money. The passengers get out and back, but it is expensive. The B767-200 had arrived in Newcastle by 1993 so the pilots flying it had an easy life. It could make anywhere in Europe effortlessly, whereas we were still dragging this steam-driven B737-200 around the track. By August 1993, I was exhausted. I had to go to Heraklion in Crete and couldn't get there and back, so I had to do tech stops both ways. We stopped in Amsterdam, and by the time we arrived in Heraklion, we had been flying for five and a half hours, with an hour on the ground. Operations in Luton said to make it to Milan Malpensa, and they would fly a crew out to meet me. The replacement captain got on in Malpensa and said: 'Message from Luton – they want you and Dave to come back on the flight deck with us.'

I was knackered and this is why sometimes I hate management. Did they want me to leave the four crew girls in Milan? We didn't have a base there. The girls had no money, no credit cards, only a uniform. We had a written understanding that we didn't position in the flight deck because the flight deck on a B737 is very uncomfortable. I phoned Luton and said 'I'm not coming back!' When we got back to Newcastle with BA the next day, I phoned someone in management who said the B737 fleet captain would not speak to me and I had to report to Luton tomorrow. I said: 'Tell him to f*** off – if he wants to speak to me, he'll find me in Newcastle.' I phoned Terry Dalgleish, who was on the union, and told him I'd just told the fleet captain to f*** off, and he said 'We're here for you, Scott. But I'm sure you'll sort it out.'

That autumn was the last B737-200 flight for the company. It did a bounce around all the bases just to say goodbye. The B757 had arrived, so that got us going again and there was a light on the horizon. In those days, if you had not flown an aircraft for five years, you had to do the full conversion course, lasting three months. I hadn't flown a B767 for just under five years. The company gave me the choice of the full course or the short course. I chose the full. It was to be at Luton and by the end of October, I was in the simulator. I flew my last flight on the B737-200 to Palma and when I got back I started the B767 course. What Britannia didn't have at that time was a B767 in Newcastle, only the B757, on which I wasn't yet qualified. Because I was not rated on a B757, I couldn't fly from Newcastle, which meant flying from places like Gatwick and Bristol. I spent that winter as if I was Luton-based, flying the B767. I was either in a hotel or at home in my own bed at night and not charging up the motorway to Newcastle, although I was still based there. The B757 was super-capable and all the tech stop hassle had gone. By 1994, I was flying the B757s and B767s, still based in Newcastle but with lots of rostered trips out of other UK bases.

I had always been supportive of BALPA. When pilots were not members, I would relate to them my near-crash in the Trislander in Loganair and the lack of someone to turn to for support. I decided it was time for me to stand for election to the Britannia company council. I didn't get enough votes the first time so tried again in 1994 and was elected. I wrote from the heart, saying that our terms and conditions were second to none but it was still up to us to fight for progress. During major negotiations with the company, the chairman would attend with other council members and we were given roster releases for any meetings. There was a monthly meeting in London and that was guaranteed time off. The union would pay for fares from Newcastle to Heathrow for the meetings. Britannia was always a strong union airline so not much would happen without us having some kind of say. One of my briefs was long haul. BALPA arranged for new members of the council to go on a negotiator's course. There were a few of us from our council and some from BA and Monarch, and we spent the weekend learning new skills – how to negotiate, what to say and what not to say. Being on the council increased my profile with both the pilots and management.

The management were not going to pick on one of the union guys without a very good reason, but I was still at war with the fleet captain because I had told him to 'f*** off', although there had been no fall-out from that. The new operations director was not a pilot, which was unusual in those days. He came to Newcastle to meet the pilots and he said: 'I hear you've had a run-in with one of my managers.' I said: 'Yes, I told him to f*** off!' He said: 'I've told him to apologise to you.' I never did get the apology.

Around this time I was offered a CRM (crew resource management) training position. Being on the council and/or training had always been a route into management in Britannia. I was never interested in management, but I was interested in training. I had never done the CMT (crew management training) course, as we called it at the time, but I was due to attend at some point. Since I had been made a tutor, the first thing to do was to sit through a course. One of the attendees was the captain I had argued with when we were holding over Luton in the snow, when I demanded we went to Manchester. On another occasion, he had argued with a first officer on the ground and cabin crew had to tell him: 'We can't open the door because the passengers can hear you!' The CMT course aimed to improve flight safety by showing a better way for pilots to work together and support each other. The syllabus included communication skills, decision-making, human information processing and physiology, and it offered 'pilot tools' to help in difficult situations. The course included analysing aircraft accidents and determining what skills could have been used to prevent them. As always, the good pilots thought it was useful – and the ones who actually needed the skills thought they already had them in bucketloads! After the two days, the awkward captain said: 'This is a great course. I think some people need it.' All the other guys were thinking: 'You! You need it!'

At the end of that week, I got a call: 'Scott, will you take over the whole programme? We'd like to be the first airline in Europe to introduce CRM to cabin crew and, more importantly, introduce joint CRM training with the pilots and cabin crew.' I hadn't even run a course!

I had to design this new course and could pick my trainers, although the head of cabin crew training had some of her trainers in mind. In those days, pilot training and cabin crew training were two completely separate empires. They were even based at different airports. I realised I must keep the cabin crew management onside because they were feeling a bit vulnerable, with a pilot coming in and taking over some cabin crew training. My title was 'chief instructor integrated crew training'. I needed to put a team together.

Members would need a natural ability to teach and communicate, but, more fundamentally, they would have to buy into bringing about a change in the authority gradient. This had to move away from the old approach, where the captain was always seen at the top, with the first officer flying along thinking: 'We might hit that mountain, but I'd better not tell him. I'm sure he's right.' Crunch.

The pilots' CRM course had already started to develop new ways of operating the jet but the subject had not been introduced to cabin crew. I looked at the pilots on the CMT team. All who remained, after a few had recently left, were good. By this time, in 1994, the whole of aviation knew about the tragic Sioux City air disaster when a DC-10 crashed. After the crash, I went with a senior cabin crew member to the US to visit United. We wanted to get a feel for what the Americans were doing in the realm of joint CRM because they were clearly ahead of us. They gave us interesting videotapes in which the captain and surviving crew members of the Sioux City crash spoke about their experiences. Al Haynes, the captain of the crashed DC-10, said the CRM had saved their lives. United had identified the need for CRM training and produced a course for captains called CLR (command leadership resource), which Al said had made all the difference to the outcome. United realised the course would benefit all flight crew so CLR was developed into CRM. 'Crew' still meant the flight deck until finally they introduced it into the cabin. Joint CRM, where all aircrew were in the same classroom, was still very new in the US. When we visited United, joint training was still SEP (safety and emergency procedure) training. CRM, or human factors training, had not been introduced into the cabin.

My idea was to have a team of half a dozen pilots and the same number of cabin crew. I had come back from Virgin more aware than ever of the importance of cabin crew, although there had always been a good relationship between pilots and cabin crew in Britannia. I went to East Midlands airport, where the cabin crew training centre was, and met the person in charge. She told me: 'There's your team.' I could see some of them were in her close circle so all would be reported back to her, but I was confident this was all going to work.

During the first half of 1995, I almost stopped flying because I was so involved in producing the course. It had to be ready for September for the winter recurrent training programme to keep the crews compliant. I did the odd flight but just to keep current. I wanted this course to be ground-breaking, slick in presentation and something the crews would find interesting and different. We decided it would be a two-day course, as it would be good to have everyone in the hotel in the evening, even if they were at their own base and would normally go home. We came up with an evening communications and team-building exercise, which would be a bit of fun. It also justified the night in the hotel for all the course members.

It was all about team-building and of course that is what we wanted them to achieve on the aeroplane: working well together, supporting one another, communicating well – including through the barrier of the flight deck door – and helping the crew to make good and, in particular, safe decisions. The idea was to have an understanding of each other's roles, which they clearly didn't. They also needed to develop the ability to actively listen. We wanted to help with decision-making and task management and explain how the brain worked. We did a bit on physiology, helping pilots and cabin crew understand the barriers which might prevent them working effectively together. We also taught them to know when it was appropriate for a captain to be more assertive or to take control, and when not to be overly assertive.

The years after the war and up to the 1980s were still the time of 'captain and boy' on the flight deck. Once, when I was a first officer over Barcelona at 37,000 feet on my way to one of the Balearic islands, ATC was giving us a very late descent due to crossing traffic below us. We were very high on our descent profile and I was diving off our height but still high. I called for 'speed brake' and there was no response. I called again and the captain said: 'We don't use speed brake in my aircraft.' I thought: 'What do I do now?' At the time of the course, we were trying to have a more flexible authority gradient between the captain and the first officer. It would change as required by the situation. We wanted captains to welcome input from their co-pilots. It was about appropriate assertiveness and authority gradient within the crew, not just the pilots.

What we were trying to get over to the first officers was that although the captain is the boss, he might be making a mistake. As first officer, it's your responsibility to stand up and do something.

The East Midlands crash of a jet on to the M1 showed clearly that the cabin crew had a great deal of information that the pilots could have used to save the jet from crashing. Cabin crew were NEVER encouraged to bring information into the flight deck. They generally thought they were not competent to tell the pilots about operational information. They might call and say there is a funny smell or a noise at the back of the aeroplane but not much more. In this case, the pilots never asked the cabin crew if they had any information, and the cabin crew thought 'We don't know what we're talking about'. The outcome was the crash and loss of life.

We had to work out an exact language of communication because talk of left and right was of no help if the other person is facing the opposite direction and most of the cabin crew didn't understand port and starboard. Aircraft hardly ever crash purely through a technical malfunction: usually, human factors are involved. Sometimes it is completely the pilots' fault, as it was with a B757 from American Airlines that hit the hills in South America. They got lost and confused during the descent through very poor situational awareness, just blindly following the magenta line to the crash site. When the EGPWS (enhanced ground proximity warning system) called out 'whoop whoop, PULL UP, PULL UP', they flew a textbook terrain avoidance manoeuvre. In those days, it didn't say to retract speed brakes and they hit the hill. Had the speed brakes been retracted, they would have just missed the ridge. Boeing then included retracting the speed brakes as part of the memory drill for terrain avoidance. Of course, not being lost in the first place was the point.

If pilots and cabin crew were trained together, they would have a better understanding of each other's roles and we demonstrated that in a simple module. We asked all the crew in the classroom to write down the scariest thing that could happen on the aeroplane. Almost all the pilots said: 'Cabin fire.' Almost all the cabin crew said: 'Engine failure.' We let them explain to each other that the cabin crew could deal with a cabin fire, and the pilots, who practised engine failures every six months in the simulator, were quite confident in dealing with that. They knew their own roles but had little or no knowledge of the job on the other side of the flight deck door. We would tell them: 'We're rolling down the runway on take-off and you suddenly see flames coming out of an engine. What are you going to do?' All the cabin crew agreed: 'We're going to push the Alert button.' Then we said: 'Right, pilots, what are you going to do?' They replied: 'Nothing.' Are you going to answer the Alert call? No. The cabin crew were very surprised and said: 'We need to tell you about the engine fire.' The usual reply was: 'The engine is always on fire! We have to concentrate on flying the jet. As soon as we have it all under control, we'll get back to you.'

The whole module could have been written on the back of a fag packet but the interaction and learning that went on was fantastic. Our MD sat in on one of the early courses, as did our CAA flight ops inspector. The CAA asked if a few of their training inspectors could come on the course, followed by the RAF asking to attend. It turned out that everyone was very pleased. We were IT in Europe - and other airlines came along, including BA. One of the things that had been introduced was the phrase: 'Captain, you must listen.'

There had been a few times when a co-pilot tried to say something but had been ignored. 'Captain, you must listen' is now a recognised phrase and, when used by a co-pilot, the captain must acknowledge the call and act to take the safe option. The co-pilot might be wrong but, if that were the case, at least the jet was safely away from the ground. On the other hand, the captain might come to his senses and declare: 'You're right, let's not land here.' After the introduction of joint CRM training, operational results were seen straight away. Within the first season after completing the course, the crews were reporting back more information than ever before. A funny noise at the back of the aeroplane, a bit of vibration, ears were popping – it all got reported to the pilots, who were happy to accept this information. Most of it would be nothing but that wasn't the point. One day, lives would be saved as a direct result of our course.

Learning from the spooks

On one of my visits to an American airline, I saw they had a whole line of their own simulators. They also had a section of an aircraft passenger cabin, with a simple flight deck at the front, for training crew evacuation procedures. On one side for land evacuation and on the other with a swimming pool for water evacuation. Like almost all other airlines, we didn't have the money for that kind of equipment but, undeterred, we ran a classroom-based role play with the pilots sitting as they would in the flight deck, the cabin crew behind at their door positions, and the rest of the class as passengers. The scenario started off airborne at cruising level and over the Atlantic on the way to Tenerife. A card was handed to one of the passengers saying that they had found something suspicious in the rack above their seat and they should tell a crew member. A bomb was found and the whole crew had to deal with the problem and achieve a safe outcome. Even in a classroom, some of the crew found the whole thing a bit scary.

We had a new procedure to be followed on finding a bomb, or IED (improvised explosive device). To prepare for the IED procedure, I had a memorable experience in a rather unexpected environment. I was sent for a week's training with MI5. We looked at IEDs in their various disguises, hijacking, and the different negotiation techniques. We visited the AAIB (Air Accidents Investigation Branch) at Farnborough, where we were shown the wreck of the Pan Am B747 that crashed at Lockerbie in Scotland. They explained how they had worked out exactly what happened to the aircraft. They also explained a number of steps we could take to mitigate an explosion on board an aircraft at altitude. We spoke to Boeing about this and they confirmed the LRBL (least risk bomb location) was in the middle of the back right-hand door in the cabin. We were shown how we should make a platform at the LRBL and place the IED on the platform. We should then fill the surrounding area with pillows, cushions, blankets and anything else soft that the crew could lay their hands on. The plan was that if the IED detonated, it would blow out the door and not compromise the aircraft structure too much.

By 1998, I was away nearly all the time at East Midlands, and I had become almost completely absorbed in the company. My life centred around the hectic activities of the workplace, and it had taken its toll. With all that going on and the union role, I decided I'd had enough of it. I'd had a bellyful of life – and I quit. I resigned from the training role and from the union.

I was then moved to Manchester and, being a Manchester-based captain, I was allowed to operate long haul on the B767-200. My first long-haul flight in command was Manchester-Santo Domingo, the capital of Dominican Republic. I took Ruth with me in the flight deck. She had a fabulous window, five times the size of a normal passenger window. Then we got the B767-300, which was the stretched version of the aircraft, capable of taking 349 passengers. The trouble with putting 349 passengers on a B767-300, plus their luggage and possibly some freight, is that you can't lift the weight off the shorter runways to get to Mexico and the further destinations. Manchester, Gatwick and Stansted have long runways but Luton, Bristol and, until fairly recently, Birmingham have shorter runways. At that time, a direct flight was dependent on the passenger load on the day. Equally important was the weather, both for take-off and en route. However, in later years, by reducing the payload by about 100 passengers and their bags, it can be done easily.

The company advertised the role of line training captain, where you teach and check pilots on the aircraft while flying the line with passengers on board. I applied, was interviewed and, with a few others, got the job. I was chuffed! I had already done the core course and in 1998 I did the simulator training and then the practice line training with another senior training captain acting the role of student. This sort of training is very demanding because the senior trainer is trying to push the normal operating boundaries, so the 'new boy' – ie. me – has to recover by possibly taking control. All of this happens with passengers on board heading off on their holidays.

That year, there was a vacancy for a captain in Birmingham. I was slowly working my way back down south to be nearer home. I could more easily drive there - about an hour from home - and it is a great base. Luton can be half an hour or two hours away, depending on motorway conditions. Birmingham had a nice mix of flying and a long-haul programme operated by the local pilots. The main long-haul bases were always Manchester and Gatwick, but Birmingham filled the gaps. It turned out to be a busy year because we had taken on a whole group of low-hour second officers and I was cutting my teeth training them.

I had joined Britannia with about 2,000 hours, which was the norm. The few low-hour pilots were hived off to an executive jet company in Luton or to other small UK airlines to gain experience for a couple of years. The second officers now joining often had less than 200 hours total flying time. They had come off the street with no flying experience and completed an 18-month full-time course, then been spat out with a frozen ATPL/IR that entitled them to be a co-pilot on any aircraft they were trained on.

By the time I flew with them, they had completed the B757 technical course and the simulator course and had been base flying with an empty aeroplane for eight to twelve take-offs and landings. The first six or eight sectors with the second officers were with a 'shotgun pilot', usually a first officer, whose role was to support me if things got busy or too much for the new pilot. I could use them to carry out the outside walk-around or look after the radio to take the pressure off the trainee. In extreme cases, I could put the trainee in the third seat and the experienced pilot in the right-hand seat.

Most of the new pilots did very well, considering their lack of flying hours, but they lacked something. They hadn't been out there on their own, scaring themselves witless. Some would say: 'That's good, because they are only going to be flying airline passengers in an exceptionally controlled and regulated industry.' I would disagree. Once you have frightened yourself in an aeroplane, you realise that your passenger jet can and will kill you, given a chance.

I went into Corfu one night while training a new second officer who had done about 10 sectors. There were just the two of us. On the way down the coast of Italy, I asked him if he had been to Corfu before. Looking at me with big eyes, he replied: 'I've never been to Corfu and I've never really flown at night.' I thought: 'If I die…!' I asked gently if he didn't have to fly at night during his initial pilot training, and he said: 'We flew during official night but it wasn't really dark.' The majority of these trainees passed and went on the line, but if it was Scott's airline, I wouldn't have 200-hour pilots on the jet. It's not fair to anyone.

BEA and BOAC, both now BA, had for years sponsored young well-qualified people to go to their pilot training college at Hamble. The new second officers would then go into the third seat of a Vanguard, a Trident or a VC-10 and would not fly the aircraft unless the weather was CAVOK (visibility more than 10km, no cloud below 5,000 feet, and no significant weather). Their role was to pick up weather reports, talk to ATC and learn the job of co-pilot.

The training was thoroughly enjoyable, with the odd challenging moment. Landing on R/W 33 at Birmingham with a westerly crosswind is challenging enough for any pilot, but there were a couple of times when we were about to give the runway a glancing blow and head off in a direction not down the runway. On those occasions, I had to grab control and firewall the engines as we lifted back into the air. I thought: 'I might not be going to kill myself but I could break a jet doing this.' Part of the problem was their low experience, together with my low experience as a training captain on a jet aircraft. I was still learning when and where it was inappropriate to let them have a go. I was also learning when to take control in such a way that their confidence was not damaged because they were screwing it up. I had been an instructor in light aircraft and some of the skills did transfer over, but it was a big learning curve for me as well.

One of the lads, who struggled with his landings, did eventually pass his final line check and went on to fly the line. Not many months later, he buried a jet on landing and was removed from the line. He was given further training but left the company shortly after. When someone is really struggling with hand-flying the jet, you might not be doing them any favours by continuously giving them the benefit of the doubt. If there is steady improvement, then fine, but if not, it is probably best for everyone to let them go.

Years later, when I was the flight crew training manager, we were again taking in some very inexperienced young pilots, with mostly good results. I used to get weekly reports from one of the senior trainers down in CTC (the training school) if a student was struggling during the simulator phase. I quickly learnt not to ignore this and just hope it would get better. I would ask the boss in CTC to take the next sim session and report back with a 'yes' or 'no'. If it was a 'no', I would have a face-to-face with them, along with a BALPA rep and HR. The outcome would be that they wouldn't be flying for us. We would write off all their training costs and the money they owed us, and give them a good reference. I usually told them: 'Go and fly for Loganair, or a similar airline, where you will be flying slower turbo jets and doing lots of sectors each day. That way, you will build up your handling skills, and then try the jet airlines.' One chap asked: 'Can I re-apply then to come back?' Of course. He hadn't done anything wrong, he just couldn't handle the jet.

When I was still running ICT, we started a long-haul airline in Germany called Britannia GMBH. We were going to employ local, or at least new, pilots to fly for this airline but in the short term our pilots could volunteer to operate out of Germany for extended periods of time. I was asked to go over with HR to be part of the interview team. While there, I noticed one captain's file in the 'no' pile. He seemed perfect for us and when I asked HR why he wasn't being interviewed, they replied that he had speeding points on his driving licence, which meant he 'had a problem with authority'. I said: 'But you've offered this other person a job who is so young he's hardly held his driving licence long enough to get any points.' The HR people sieving through this list had no idea what a good pilot looked like.

They were mostly German pilots, and some Dutch. We had British B767-300s based in Germany flying to the Caribbean long haul out of Hanover, Frankfurt and Dusseldorf. Initially, they were crewed by British pilots but the locals were quickly coming on line. This was the first time anything like this had happened – a British airline starting a German carrier - and it didn't go unnoticed by the other big German companies. By 1998, I was a line trainer and quickly got involved with the German training programme. I was doing both first officer line training and captain line training. The new pilots were all experienced and mostly B767-qualified so it was usually easier than with inexperienced second officers.

The airline was doing well, profits up and confidence high. Then suddenly there was news that the company was about to be bought. We had been floated on the stock market and bought by a German company. It turned out that the directors who had come in and given the airline a good kicking a couple of years before had done their job and upped the profits, certainly in the short term. The days of pilot MDs had almost gone and, in our case sadly, we became a stronger airline for it.

Whizzing off to Oz

My flights to Australia began in March 1996 when I was still on the union. In fact, I was responsible for the long-haul bid in those days, whereby eligible pilots bid for the long haul they wanted to be involved with. A bid would be placed and it would be my responsibility on behalf of the union to make certain, when the bid committee sat to allocate the trips, that everything was done properly. By this time, I had flown quite a few flights across the Atlantic but none eastbound. The Australia flights started in the winter of 1988/89 after I had written the ETOPS course and had then gone back on the B737-200.

The Australian flights seriously undercut the big airlines on the route, but at that time we were flying the B767-200 which didn't have the legs of the later 300, and certainly not the range of the B747. We needed two stops to get to Australia, whereas the B747 could do it with one. On the plus side, the seat price was low. If you went with BA or Qantas, you would be shelling out some £1,500 while our seats were coming in at £470. Because we were a charter operation, we also had to include one night in a hotel in Australia. The flights were all full, even though it might have felt like it was taking a fortnight to get there. They didn't care as they were getting tremendous value.

Our B767-200s carried 274 passengers, all in Economy in those days with high density seating. All the flights tech-stopped for fuel and a crew change in Abu Dhabi and then in Singapore. Abu Dhabi was like a Britannia base. We had the flights through to Oz and also to India, Sri Lanka and the Maldives. We stayed at the Intercontinental in Abu Dhabi, which was a lovely hotel, and they gave us two connecting rooms we could use as our permanent crew room. Every time I passed through there, the crew party was in full swing. Crew arrived and crew departed, going in both directions, but the party never stopped. Even Ramadan didn't stop it.

I put in a bid for a flight in March 1996: Gatwick-Abu Dhabi-Singapore-Sydney. We positioned across to Melbourne for a few days, then operated Melbourne-Singapore, Singapore-Cairns and then Cairns-Singapore-Abu Dhabi and Abu Dhabi-Gatwick. The only place on this trip I had ever been to in my life was Gatwick!

It was a completely new experience for me. The first officer, Rod, was a bit older than me and had been a Vulcan captain in the RAF. He was a very interesting character and I enjoyed his company. We got to Gatwick and a training captain advised me: 'Take an extra tonne of fuel.' So I did, and off we went for Abu Dhabi. There were no take-off performance problems because all the sectors around the route were fairly short at around the 7 hours mark. The hardest part of the first sector to Abu Dhabi was crossing the group of countries between Cyprus and Saudi. We went down to Cyprus, then north of Israel over Lebanon. Routing could vary. Then we were into Jordan and Saudi and on each border they required a 10-minute early call. We tended to work two frequencies – the one we were currently operating on and the other to speak to the next country, or even the one after. I used to ask myself: 'Don't these countries have phone lines so air traffic can co-ordinate with each other? Obviously not!'

I did that run many times over the years. I would be away each time for about three weeks. The flying time to Abu Dhabi was 7 hours 10 minutes, and we stayed three nights. On the first night, the cabin manager asked if I had been to Australia before and I said: 'No, this is my first trip.' She then asked if I had ever been to Singapore and again I said: 'No.' So she said: 'You won't have had tiffin at Raffles Hotel. I will book it for us all on the first clear day in Singapore.' What a lovely cabin manager and what a treat it turned out to be. We flew down to Singapore, crossing the whole of India where the view was incredibly clear that day. We saw the Taj Mahal from 36,000 feet. A fantastic sight – and I was getting paid for this! After a clear day off in Singapore and the visit to Raffles, we flew on to Sydney, where we spent a night and a day. We went out to a lovely waterside restaurant for a drink. The next day, we all had to position as passengers to Melbourne with Qantas – and one of the girls climbed into the crew transport with a headscarf pulled over her forehead and in very dark glasses. She was not a well cookie.

At the Hilton in Melbourne, I noticed all these chequered flags in the hotel lobby. The Australian Grand Prix was the next day - just across the road. This was the first year the race had been held in Melbourne and we were there! Rod and I, plus one of the stewards, watched the race. It was my first Grand Prix and it was great fun. The noise was unbelievable, even with earplugs in. At the end they let the crowd on to the track and we wandered around the circuit, then into town where the whole place seemed to be one big party.

We left Melbourne for Singapore, and a couple of days later, flew to Cairns and suddenly we found a very different climate to Melbourne, which was hot and humid like Singapore. We arrived in the morning and were due to fly back to Singapore that evening. On the flight over, the crew were making plans to see the Great Barrier Reef. Due to the short time there, they decided to fly out on a float plane, then spend time snorkelling. I didn't want to feel knackered on the return flight , so I went to my room instead and switched on the local TV. It was there that I heard about the Dunblane massacre in Scotland and I felt quite strange and very far away. It is a moment I always remember. Here we all were in a kind of never-never land and this awful tragedy was happening in our own country. I was with a good crew but it still made me quite depressed. By the time we reported in the evening, I hadn't used my food allowance, so I wandered into the hotel shop and used up my allowance by buying a boomerang. The crew arrived looking different shades of bright red. Rod positively glowed. They had been out on the reef snorkelling and were well and truly burned.

It was a great trip, and I enjoyed flying with Rod and listening to his stories about flying the Vulcan. It was pretty clear that the Vulcan crew's mission of being ready to deliver nuclear warheads was based on the simple premise that if they made it back home, they would be lucky. In those days, they were advised that if they did complete their mission, there probably wouldn't be much to come back to. Clearly, flying the Vulcan was nothing like going down to Palma and back. Anyway, he survived all of that.

During the summer of 1996, I went to the usual destinations in the Caribbean - San Juan, Punta Cana, Puerto Plata, Santo Domingo and, of course, Orlando. I was still Manchester-based but bidding at each base bidding round to get further south. The following winter, I did another Australian series of flights. I used to try to bid for one a year. The second trip was a bit different. Britannia wanted to start doing round-the-world trips with Thomson Holidays: Abu Dhabi-Perth, then to somewhere like New Zealand, across the Pacific to LA, and onward to the UK. We were now operating B767-300s as well as the 200s. The 300s arrived in 1996 and were longer so could carry more passengers. They had the same wing, different high lift devices and much bigger engines. They gave us a bigger range although we were still high density, meaning we couldn't fully utilise all the extra range as we were carrying 348 passengers. Later, that same flight carried only 258 passengers.

Before the round-the-world route started, they were trialling the Abu Dhabi to Perth sector with a lower passenger load and no freight, flying for 11 hours-plus. I operated the first rotation with a heavy crew. A third pilot joined us in Abu Dhabi for the two long sectors to Perth and back. We would have just 12 hours off in Perth. We had a crew rest area, which was basically a curtain around six seats on the left of the cabin over the wing area. It was not ideal, being far away from the flight deck. We subtracted an hour for the take-off and climb and an hour for the descent and briefing, then split the remaining flight time into three, each of us taking three hours out of the flight deck or resting on one of the jump seats. The usual way of deciding who had what break was the pilot flying chose first, then the pilot monitoring - and the third pilot took what was left.

The flight out was uneventful, save for a few seconds when I was on my break in the rest area and the two first officers were up the front. They ran into the only cloud in the sky and there was a sudden jolt like hitting a speed bump too fast. I immediately woke from my doze, stuck my head up and thought: 'I can't do much from here, best to just trust them.' I took my rest on the return flight in the flight deck, all the same.

One of the factors which control how much load you can lift on take-off is air temperature. The hotter the outside temperature, the less power the engines produce and the less you can lift. We were due to leave Perth early in the morning before it got hot and with only passengers and no freight.

When we arrived at the jet, I was informed we were carrying five tons of plums. I said: 'We're not supposed to have any freight!' We worked through the take-off calculations and discovered we could just about lift the weight as long as the temperature didn't go up by more than a couple of degrees. By the time we were nearly boarded, the temperature was up four degrees and, after recalculating our take-off performance, we were a tonne overweight. I said: 'Unload a tonne of plums from the aircraft.' They asked where they should put them and I said: 'On the grass!' I had been waiting to launch and the day was getting hotter, and my ability to lift the load was getting worse by the minute. I finally said: 'Take three tonnes off and we are going. If we don't get off in the next 10 minutes, you can take the lot off.' Things don't really change.

Later I was going to Cancun in Mexico out of Birmingham, just before they opened the runway extension. Again, I was struggling with take-off performance. The runway in use was R/W 15 and it was wet, air pressure was very low and it all added up to me not being able to lift the fuel required. I phoned ATC and requested the reverse runway direction (33). Even with a slight tailwind, that gave me a higher take-off weight. The weather over the Atlantic was horrid and the flight time was just under 12 hours, which meant the fuel required was high, well over 60 tonnes. If I fuelled up to the required level and the conditions worsened, I would be stuck and de-fuelling would take forever. The other option was to put 10 tonnes on and fly to Manchester, where the runway is longer, and change the crew because we would be out of duty hours. Then the new pilots could take the delayed flight across the ocean. I phoned our Operations and gave them the options. They hummed and hawed and asked what did I think? I thought: 'It's decision time.' So I said: 'I'm going to put the fuel on and have a go.' We taxied out. The conditions didn't change and we launched on the direct flight to Cancun. The block time was 11 hours, 50 minutes and we made it. Sometimes someone has to make the call and that time it worked well.

I did another long-haul flight with Rod as my first officer in 1998, this time to New Zealand. It was January and we did Manchester-Abu Dhabi, then a few days later Abu Dhabi-Batam, an Indonesian island just across the Straits of Singapore. During our few days in Abu Dhabi, we were invited to a party by Tim Hiles, who was a training captain when I was in Virgin. He was now boss of the 'special flight' in Abu Dhabi for the Emir.

The party was held on a penthouse floor in downtown Abu Dhabi. Everyone there was crew, from all corners of the world. Interestingly, no one really wanted to be there. They were just working for the money. They used to operate an old Jumbo SP (Special Performance) which had a bathroom area decked out in granite and huge chandeliers. The aircraft was so heavy that they couldn't go very far because they couldn't lift the fuel. They had an anti-missile suite at the back of the aircraft where they could take defensive measures. Tim, who was not only captain but the boss of this operation, was not allowed downstairs during the flight. Behind them would fly an Airbus carrying all the bags and stuff they couldn't get on board. They had white leather suites in the Jumbo and, because the Emir's family loved their birds of prey, these birds would also be on board, crapping on the furniture.

I enjoyed passing through these countries but the thought of working there full-time would be a completely different thing. You fill a role that the locals don't want to do and, in return, you are paid reasonably well - but put a foot wrong or think about a union and you are out. I enjoyed going to the souk and wandering around, but most people working in the hotels were not local, the doctors and nurses were not local. The whole infrastructure was imported.

We departed to Batam, then on to Melbourne and, a few days later, we positioned out to Auckland with Qantas, as passengers. We were there for a few days so a few of us hired a car and drove off to Rotorua, an area of forests and lakes with hot springs and the well-known Pohutu Geyser. The countryside we drove through was very similar to the Highlands of Scotland. Once you got out of Auckland, it was very much like being back in Scotland in the 70s. From Auckland, we flew to Batam, a long flight passing over Lord Howe Island, an irregular crescent-shaped volcanic remnant in the Tasman Sea which from altitude looked like a huge sail on a schooner. We crossed over the middle of Australia and were lucky to see Uluru through a large hole in the clouds.

So that was my third and last Aus trip – Britannia stopped those flights that winter. The big players had dropped their prices and bookings showed that passengers preferred to stop only once in each direction. We never did start the round-the-world flights from the UK. They would have needed to take a lot of seats out to give the jets the legs to cross the Pacific, so tickets would have been too expensive for Brits.

I'm on the telly

On a long-haul flight out of Manchester to Orlando, I found myself on TV. Britannia was featured on a programme called Airline – a fly-on-the-wall look at the daily operations from within the industry. The programme went on to look at easyJet for years afterwards but Britannia did the first one and it was quite a success. The TV company wanted to do more but Britannia eventually said: 'No, we've just about got away with the first one without looking stupid. Let's not risk it again.'

I was doing Manchester-Orlando in July 1997 and as I taxied out from Manchester in a B767-300, I had an IGD (integrated generator drive) failure on the right engine. That meant the only working generator was on the left engine and we would have to run the auxiliary power unit for the entire flight across the Atlantic with no standby if that failed. I disconnected the drive so it didn't eat itself and taxied back to a remote stand so as not to monopolise the air bridge at the terminal. I decided to keep the passengers on board to save the time and hassle of coaching them to the terminal and then back again. I told them that the engineers were on the way out to look at the generator drive.

Someone arrived wearing a Britannia uniform, skipped up the steps and said: 'The TV people have heard about this and would like to come on board. Is that okay with you?' I thought: 'I'll just shove off now and leave them to it. It had been agreed that it was up to the captain in all circumstances to decide if the TV crews would be allowed to come on board. When I saw the van approaching, I wandered down the steps and around to the right engine to chat to the engineers. After 20 minutes, the van which brought them over had gone. As I walked up the front steps, I walked straight into a camera coming out of the front left door. They began asking me about what was happening, what the delay was and what I was doing about it. They said: 'Any chance of coming on to the flight deck and filming that?' I said: 'Okay. Just sit there and say nothing.' They were actually pretty good. We were delayed for two hours and in that time they'd been interviewing the passengers. Put them in front of a TV camera and it seemed they loved it.

While they were filming, they had got out of me that my daughter Inga was doing her driving test that day. It was time to get going again and they were standing just clear to the left of the jet, filming us starting engines. As they were doing that, a call came through on my mobile to say that Inga had passed her test, so I leaned out of the flight deck window and yelled: 'She's passed!' I thought no more about it and off we went.

A week or so later, I launched for Mombasa, after which I had the unfortunate encounter with the stroppy passenger who demanded access to a private jet when I diverted to Dar es Salaam. At least the film crew hadn't got on the flight and thank goodness they missed all the drama. They had been at the airport hoping to join me on my trip but the aircraft was full. Not many weeks after my return to the UK, I had an unexpected call from the BBC inviting me to a TV awards event. I knew I had come over quite well and they wanted to use me in another spot.

I realised that even after a 30-second spot on TV, I would become recognisable to the general public, and, more importantly, to my passengers. If I was ever going to have to speak firmly to an awkward passenger, being recognised as the pilot on the telly wasn't going to help, so I said: 'No, thank you.' When it was eventually shown on TV, I was made out to be in a B757, not a B767. I was operating a flight to Europe, not the US, and I had a passenger problem, not an engine generator failure. There wasn't much left they hadn't changed. And they made a big thing about my daughter and how she had been taking her driving test that day. My little clip has been shown dozens of times. I have even switched on the TV in Orlando and seen myself talking about Inga's driving test. People still come up to me and say: 'You're that pilot on the television.'

In 2003, the year I flew on the Hajj, I operated one of the positioning flights to Abu Dhabi and the next day on to Ujung Pandang. I was a training captain conducting an ETOPS clearance flight on a captain on the sector to Ujung Pandang. All the crews needed to be ETOPS-cleared because it was sometimes necessary for the sector crossing the Indian Ocean. There were three pilots on board inbound to Ujung Pandang and none of us had been there before. We all studied the company airfield brief and chatted about the many different approach variations. In the end, the weather was benign and we flew a straightforward ILS approach. The terrain around the airport and the tropical weather make Ujung Pandang and similar airports in the region demanding. Add in local ATC, and the outcome is very demanding. You could be sitting in your plane, on the ground or on the approach, and the rain would be hammering down. You would ask ATC for the weather and they would read the report they had received 20 minutes before, with no mention of rain. We would say 'Look out of the window - is it raining?' and they would say: 'Yes, very heavy!'

We spent the night in a hotel in Ujung Pandang, then flew to Jeddah, via Batam. Flight times between Batam and Jeddah were usually 10 or 11 hours. All the passengers and cabin crew were Indonesian, and many of the passengers had never seen an aeroplane close up, far less been in one. Whole villages clubbed together to send one or two people on the Hajj. It was a very special experience for many of us. When we changed pilots in Batam or Singapore, they would line up to shake our hands and thank us. You don't get that on a Palma flight.

We were actually flying for Garuda Indonesia Airlines and using their call sign. The majority were not full-time cabin crew, but volunteers from different walks of life to fly the Hajj. It was seen as an honour to be trained as cabin crew. There were some language barriers and only the senior crew member would ever venture into the flight deck, but with the odd thumbs-up and smile we got through the flights surprisingly well.

Many years before we operated the Hajj, an aircraft had a cabin fire because passengers had lit a primus stove inside. It was the old world colliding with the new. We always carried a doctor though they believed that if you died on the way to or coming back from the Hajj, it was a great honour.

We didn't get involved too much with the passengers, who couldn't speak English. During the flight to Jeddah, the crew would sometimes ask when it was dawn in Mecca, so we would pull out our tables and try to work out local time. We usually came up with different answers and would take an average. I found Jeddah menacing and generally unpleasant. I thought they might make us welcome because we were flying in the pilgrims, but they didn't. On arrival at the airport, they took our passports and we didn't see them again until just before departure. We were taken to a rather nice hotel but were advised not to go out for a walk. The whole experience was weird on several levels. I checked in to find I had a suite with seven rooms and a fountain. After wandering around, looking into the various rooms, I felt I was locked up in my own prison.

If we went out for dinner and were with one of our female first officers, the whole process was unpleasant. The women had to dress from head to foot in an all-black garment, and when we went to a restaurant they couldn't eat with the men. So we would be put in a private room with no locals. It spoilt the evening, so we tended to stay in the hotel. In the past, we would have had a whole crew in Jeddah and the place was generally less aggressive. It was more acceptable to wander down to the souk, but it had been known for the religious police to usher westerners, including our crew members, to watch a public beheading or the loss of a hand. Getting a hand chopped off in front of the infidels is regarded as pretty bad - and I guess getting your head chopped off is even worse. After a few rotations through Jeddah, the company said: 'Stay in the hotel. Eat what you like and we'll pay, but stay there.' It was good advice. I went out for a walk once close to the hotel just before the 'stay in the hotel' advice and got spat at by a passing driver, so I thought: 'I'll stay put!'

On my next flight out of Singapore to Jeddah, the toilets started to block and wouldn't flush. With hours still to go, we thought we might need to stop somewhere and have the toilets serviced, until the Indonesian cabin manager came in to the flight deck and said: 'It is all right, captain, keep going. We have put notices up saying no pooing – only weeing!' So we carried on and made Jeddah – this aircraft was not stopping for anything. When we had a couple of deaths on board, we didn't stop. The crew didn't even tell us until we landed.

The jet that roared

When I was Birmingham-based for the first time, I was flying with a young first officer who was an artist and had paid the best part of £70,000 to gain his licence. He had a studio in New York but after painting portraits of people who said 'I don't look like that!', he thought: 'I don't want to do this for the rest of my life.' He decided that a pilot's life was for him.

In August 2002, he and I were flying a B767-300 to the Canary Islands but we didn't get there. We were bound for Tenerife and were just reaching our cruising level, heading out over the Isle of Wight. We were still working London ATC as we approached the French boundary when the back of the aeroplane called up and said with some alarm: 'There is smoke in the rear galley.' I asked what they were doing and they replied: 'We have discharged some BCF (fire extinguisher) into one of the ovens.' I told them to keep me in the picture. It does occasionally happen that old food left in an oven can burn or even catch fire, and the first thing the crew will notice is smoke seeping out past the oven seals. It is not usually any problem. The oven will contain any fire and once the oxygen is used up, it will extinguish. The crew are trained to crack open the oven door and discharge a small burst of BCF into the oven and close it. Any fire would immediately extinguish.

Seconds later, they phoned again and announced: 'We've got a lot of smoke coming out from behind the ovens behind the panelling around them. We are using BCF extinguishers and using a fire axe to get behind the panels.' This was a whole new ball game. At that point, I told Ian to declare a Pan call - an urgency call. While I was off the radio talking to the cabin crew at the back of the aircraft, London ATC had handed us over to Brest Control. So Ian transmitted: 'Pan Pan, Pan Pan, Pan Pan, this is Britannia...' Their reply was: 'Say again!' The French are normally very good but I think they were just taken aback. My next order was: 'Upgrade.' We declared a Mayday.

'Mayday, Mayday, Mayday, this is Britannia... ' Immediately they asked: 'What are your intentions?' I said we had fire in the cabin and wanted immediate descent and to divert to Gatwick, and they told me to change to London Control. I called up London on a Mayday and said: 'I have smoke in the cabin and I want immediate descent into Gatwick.' ATC acknowledged and cleared us direct for Mayfield, which is a navigation beacon close to Gatwick. I said: 'No, I need to go straight to Gatwick. I want you to radar vector me to a tight right base for Runway 08.' They fully understood and gave us immediate descent and a heading for Gatwick, and cleared everything out of the way. My job was to get the jet on the ground as soon as possible, preferably on to an actual runway. The cabin crew's job was to tackle any fire and control the passengers. I put the aircraft into a high-speed dive and aimed for the airport.

I pushed the 'Alert' button, which sent three chimes throughout the cabin and brought all the crew to an alert status. I was diving and accelerating up to MMO/VMO - the fastest Mach number and subsequently the fastest indicated speed that the aircraft is certified to fly. This speed is known as the 'barber's pole' because it is indicated by a coloured needle that looks like a barber's pole on the air speed indicator. My rate of descent was off the clock, in excess of 6,000 feet per minute, more than twice the normal rate of descent. From 37,000 feet we would require about 120 nautical miles to touch down. I was a lot closer to Gatwick than that and even with my high rate of descent, getting the height off was still a problem. I wanted to get below 10,000 feet as quickly as possible and then depressurise the aircraft. We were not getting any reassurance from the back, so the cabin manager went down to investigate and take charge. She called again as we were diving through 15,000 feet to say there was still smoke coming from behind panelling and they were trying to get the panels off.

There have been a few aircraft cabin fires over the years and experience suggests that once the fire takes hold, there is very little time before it becomes catastrophic. Within 10 minutes, a fire can burn through flying controls, hydraulic lines and oxygen tanks, and, unlike iron and steel, aluminium burns violently in air. Structural failure can happen quickly. A pilot's nightmare is an uncontained cabin fire over the ocean, where there might be little time available to carry out a controlled ditching before the onset of structural failure. At night, a controlled ditching would be extremely difficult to accomplish. Burning internal cabin furnishings produce toxic smoke, which would kill everybody very quickly unless passengers were on 100% pressurised oxygen - and the only people on a jet aircraft who have 100% pressurised oxygen are the pilots. In the 1980s, an American DC-9 with a fire in the hold landed as soon as it could but half the passengers died. A Saudi Tristar landing with an internal fire successfully stopped on the runway but the captain didn't order an evacuation and all on board died from the toxic fumes. I have never met one airline pilot who wouldn't say their nightmare scenario would be a major cabin fire.

I knew that getting us on the ground was everything. On the other hand, there was no point in going so fast that I couldn't slow in time to actually land. This was no time for a go-around. With the aircraft under control, I asked for the QRH 'Smoke, Fire or Fumes' checklist. Ian started to work his way through the checklist and after switching off a few systems, including the utility busses which removed all power from the galleys and therefore all the ovens, the procedure then referred us to the 'Smoke or Fumes Removal' checklist. I had expected this and it was why I was so keen to get below 10,000 feet as I knew the checklist required us to be below 9,500 feet.

London ATC were great and kept giving us further descent as they cleared aircraft out of our way. The smoke removal checklist required us to open the cabin air outflow valve at the rear of the aircraft. This would suck any smoke and fumes out the back. It would also completely depressurise the aircraft. As we descended through 10,000 feet, I told Ian to depressurise. Then something happened that I'd never heard about nor seen written anywhere, and neither could we have demonstrated it in the simulator. The noise level inside the aircraft rose exponentially and turned rapidly to a screaming roar because we were diving so fast. All the rubber seals around the doors had cracked open because there was no longer any pressure holding them shut. The crew became incredibly alarmed and so were the passengers, although they were strapped in and had no say in the matter. We needed the crew and at that point we had lost effective communication. They couldn't talk to each other from front to back with the handsets because they were useless due to the incredible noise.

Ian and I were okay because we had headsets on and were at the front and behind the flight deck door, but the cabin crew had a whole new communications problem. I had chosen not to put my oxygen mask on because they are incredibly clumsy when it comes to general communications. Wearing a full mask and breathing pure oxygen under pressure make you sound like Darth Vader when you talk. Had there been even a whiff of smoke at the front, I might have put one of us on to it - if necessary, both of us. It is one of the things the QRH would say one 'should' do but it doesn't say 'must' do. You need to think about the decision and I was happy with mine.

So there was this phenomenal noise and I was trying to talk to the back of the aircraft. At that point, they were reporting that they thought the smoke was dissipating. We have a bird strike maximum penetration speed of 313 knots below 8,000 feet but, given the circumstances, I wasn't too worried about hitting a wee sparrow. Nonetheless, I was starting to slow down as we were getting close to Gatwick and I had to configure the aircraft for the approach. The cabin manager called to say the smoke was clearing, which was good but that might have been due to the open outflow valve sucking the smoke out. I left it to the last minute before I raised the nose and dropped the gear and, as the speed decayed, I increased flap stage by stage until we were at a slow enough landing speed. There was no point in running off the end of the runway still doing 50 knots.

During the descent, the London controller said: 'Do you want me to put you on the ground?' Normally, you would change ATC controller frequencies a few times when descending through London's airspace and then change frequencies again to another controller or two approaching Gatwick. By staying with the current controller, it was going to cut down our workload. Most pilots don't know they can do that and I said: 'Yes.' The fact is that one controller can almost take you throughout the whole of Europe if needs be. So if you have a major problem, you can build up a relationship with that individual and they can take you the whole way. I used to do this in the simulator, when I would say 'Would you like me to put you on the ground?' and the guys in the sim would reply: 'Oh yes, that would save frequency changing.' Of course, the crew have to remember the controller is miles away. After landing, there wasn't much point in asking: 'Can you see any smoke?'

Thankfully, the landing was uneventful and we came off the runway using the high speed turnoff. We stopped and immediately the fire engines were around us. We cut the engines and, as the airport authorities had arranged for steps to be out on the taxiway, firemen were up the back steps within seconds with their heat-seeking thermal devices. They came on board and pulled bits off the panels at the back, and found the smoke was behind one of the ovens. Wiring had short-circuited there and caused an electrical fault. They took the charred oven away and, after a reassuring PA to the passengers, we started the engines and taxied in. I knew that was going to be the end of our day. We wouldn't have the flight duty hours to refuel, replan and get to Tenerife and back to the UK. More importantly, after what we had all gone through, flying again without a thorough debrief and a wind-down was not going to happen.

I knew it would take a while to re-crew the flight so we would need to look after the passengers. We were in Gatwick - a big airport with lots of crew - and I thought: 'Someone else can take this sodding jet to Tenerife.' Just at that moment, someone came out to the jet in a car and declared: 'We can't get your passengers off. There is no room in the terminal.' So we all had to just sit there for the next few hours. I was pretty hacked off that an airport the size of Gatwick couldn't accommodate a few hundred transit passengers.

Despite all this, it was one of those times when I thought: 'I quite enjoyed that. And we all did a good job.'

The passengers, not unnaturally, started getting a bit grumpy and I was asking our operations people: 'Are we going to re-crew it? Are we changing jets?' I couldn't get a proper decision, but I did say to the passengers: 'I'm trying to get you into the terminal where at least you will have a bit of space. But they say there is no room. I know, it sounds like madness.'

Eventually, I managed to persuade a senior manager to come out to the aircraft and he said he'd get something sorted. Not long after that, we were re-crewed and I could tell my passengers: 'You've got a new crew who have got enough hours to take you. Best of luck!' We went across to the Gatwick crew room, where we organised tea, coffee and sandwiches because I needed to debrief.

We sat around the table and it quickly became clear that although Ian and I had been aware of what was going on and had a reasonable idea of the overall picture, we were in a minority. Most of the crew were not sure what was going on and there had been a complete breakdown of communications because of the noise of the airflow around the jet. Someone at Door 1 said: 'I had no idea what was going on, I couldn't ask.' They started thinking the worst as the noise grew louder. I knew why that had happened but they clearly didn't, and some of them said they were really scared. It was another reminder to me that pilots know, to a great extent, what is happening in the jet most of the time but the cabin crew on the other side of the locked flight deck door seldom do.

What are we going to do now?

I learned a great deal from that episode involving the galley fire. The incredible noise level when we depressurised the jet was not mentioned anywhere in the manuals. In the simulator, we did not practise failures that ended in death and destruction. That would just frighten the crew. Everything done in the simulator has to have an achievable outcome. Multiple failures would be managed with possibly multiple QRH checklists, but failing systems until the aircraft was unflyable would be negative and not difficult to achieve. Boeing provides a double-engine failure checklist but if the crew were doing the checklist correctly, the trainer would enable one engine to re-light. If they didn't follow the correct procedures, the examiner would freeze the sim before the inevitable crash, partly because the simulator's visual picture is not good enough to try a dead stick landing unless it is on to a runway, but mainly because the pilots are completely absorbed in flying the jet and the crash could easily be traumatic and put a stop to the session. Pilots have to fly the simulator as the real aircraft, otherwise it is just a fancy computer game.

Even before I became a captain, I knew that. Sometimes things would happen where the outcome was going to be ugly. Some things happen beyond the scope of the emergency checklist. The loss of the middle engine and subsequent loss of all three hydraulic systems in the Sioux City DC-10 crash is a good example of what people thought could never happen. And, of course, Lady Luck played her part. If you lose all engines and are planning a dead stick landing, are you over the Himalayas or over the ocean at night? Is the cloud all the way down to the ground? If so, best of luck!

Sometimes when I spoke to less experienced co-pilots about situations like this, it became clear that they never got that far in thinking about real crisis situations. Of course, that's part of the reason they were not yet captains. It doesn't just happen. You have to think about problems and the best solution possible, including the loss of the jet, but hopefully not all on board. It's not just about the more usual failures we practise in the simulator, such as engine failure. When I was in the right-hand seat and coming up for command, old captains would say: 'Now, think of more ugly things, like you've just taken off from Palma and you are heading out to sea and suddenly you get a call from the rear that the aircraft is being hijacked.'

In those days, we didn't have an armoured door but we could still lock the flight deck door. You would think: 'What happens if there's a guy out there and the cabin crew are saying he has a knife at one of their throats and if I don't open the door, they're going to kill her? What am I going to do?' Years ago, the advice was that if we were on the ground, we should escape through the flight deck windows using the escape rope, thus grounding the jet. None of us liked that idea and it was changed to 'stick with the crew and passengers'. Then the question arose: What about a hijack during the cruise and people are being killed? Would we depressurise?'

I once posed the question to my first officer: 'If we had a double engine failure now, where would you go?' The chap replied that he wasn't sure, as we were in the middle of the ocean. So I asked him which way the waves come - and he said he wasn't sure. I said: 'They come from America. They come from the west. So we want to turn, to ditch in a northerly or southerly direction.'

Boeing has never ditched a B767 or any other airliner but the flight crew training manual does give some advice. I asked: 'What pitch angle do they say we should hit the water at?' Not surprisingly, this chap muttered: 'I'm not sure.' I told him the answer is 10-12 degrees nose up. He knew to touch down as slowly as possible but didn't know it had to be at the suggested rate of 200 to 300fpm. And he didn't know to touch down on the windward side and parallel to the waves. I told him: 'If you're in the middle of the ocean and there's nowhere near, you might as well go left if you're heading over to America because you will be 100 miles closer to warmer water than you will be going 100 miles right, where it's colder water.'

Clearly, he had not thought about ditching and was quite happy to fly over oceans oblivious to the remote fact that both engines might stop. At some point during the 1980s, a BA Jumbo was down Jakarta way flying into volcanic dust which hadn't been notified. The pilots could smell an acrid odour and started to see a static discharge around the windshield. It was at night - and the next minute, all four engines ran down. The impossible had happened.

The captain spoke to the passengers and said: 'We're descending and we've got a problem with the engines, but we're doing our best to get them going again.' And fortunately they did. They got two or three going and limped into somewhere safely. Since then, we have included entry into volcanic ash in simulator training and Boeing has a 'volcanic ash' checklist in the QRH.

In the examiner's chair

In the late 1990s, I was at still at Birmingham line-training a constant flow of new first officers and second officers, as well as the new German pilots. I was happy at Birmingham, but in 1999 we discovered that Ruth had cancer and needed surgery. The company asked if I would like to move base to Luton to be closer to home. I continued to fly and train, and slowly we came out of that dark place.

German company Preussag AG (TUI group), a large mining company which already owned shipping company Hapag-Lloyd and an airline, acquired the whole of Thomson Travel Group in 2000. I got the job of TRI (type rating instructor) and became an instructor in the simulator. The TRI course was in-house and the biggest learning curve was working the instructor's panel in the simulator and having the spare capacity to keep an eye on the students. As a TRI, I was mostly teaching pilots coming on to a jet from another aircraft type, or training new pilots into the company on to the B757/767. We had got rid of all the old B737-200s so there were no pilots changing type and no new conversion courses. Not long after I qualified, the flow of new students had dried up because of an industry downturn and pilot recruitment stopped. There just wasn't much training going on, so I got a part-time TRI job on the side, working for the GECAT organisation, providing simulator training and checking near Heathrow. They paid me by the hour and I ended up teaching in the old Britannia B767 simulator, which had been in our training centre at Luton.

What that showed me was that the standard of third-party training was not good. I was asked if I could go and do a few sessions with an airline, usually a foreign carrier. I didn't know their SOPs (standard operating procedures) so all I could do was basic instruction from the operations manual. I didn't know their calls and responses and, frankly, for a time I felt a bit of a fraud. The simulator company said one of their trainers would watch me training a detail in the sim and give me feedback, but they never did. Then I did some training with a former Eastern bloc company and I found out on the day that I was training through an interpreter. I thought: 'This is dreadful!' It was good learning for me but the product was very poor. I was earning a bit of extra money but more importantly, it was getting me into the simulator and I was learning the finer points of operating it and building up my skills as an instructor.

Then the company advertised for TREs (type rating examiners). As a TRE, you are at the top of your game, examining on behalf of the company and the CAA. I applied for the job, which involved a large salary increase. The course was mostly based around the simulator, and we were learning how to examine and to follow all the CAA rules on conducting tests. The whole process took months because after the week's ground school we had to run live simulator sessions under the supervision of a TRE. Then we were put forward for the final test with a CAA training inspector.

For 99% of all pilots, the highest thing they could achieve would be TRE. As type rating examiner, I had reached the top of my profession and I was proud to have achieved that. I really enjoyed going in and examining. Most trainers say they like the teaching part but I enjoyed the examining because I liked getting into the candidate's head and trying to understand their thought processes. I wanted to establish that they could manage their workload and that their situational awareness was well ahead of the jet. If it wasn't, I would suggest they slow the whole thing down until they caught up. I wanted them to manage their pace and to provide an environment where they could communicate well, and make and share well thought-through plans. That is what 'good' looks like to me. If, then, they are a few knots fast, it is almost irrelevant. The jet is safe and the outcome is assured.

I also enjoyed using and developing my facilitation skills, especially during the debriefing phase, by asking questions like 'How you did feel at that point?' I would often end the de-brief by saying: 'Tell me two points that you are taking back to the jet.' If they looked at me blankly, I had a bit more work to do. It was all about getting the learning on to the jet. I knew I had to build up my abilities in this field and I thoroughly enjoyed the process. I spent a lot of time in the simulator, especially in the winter, when the flying programme was not so busy.

I was proud to have reached this position but there was no thought in my mind about seeking a management role. I wanted to be an examiner - it was the training world that interested me and that was all. Unbelievably, I had achieved that. It was more than I ever thought I would achieve in my wildest dreams. I was spending quite a bit of time away but only in Gatwick. About a year later, the head of training asked me to be chief training captain on the B757/767. I had not long become an examiner so I said in my usual straightforward manner: 'No.' He added: 'If you don't take the job, we might need to skip a generation.' Again I turned him down.

There was a new approach to checking and training and we were at the leading edge of that. Flying five knots fast or two knots slow was not crashing jets. It was poor decision-making, monitoring and situational awareness that was causing the loss of aircraft. We had always been in the vanguard of human factors training but some of the older TREs were still concentrating on examining 'by numbers' and debriefing by going through a list of mistakes in chronological order, seen over the previous four hours in the simulator. TREs needed training in the benefits of crew resource management and we newer examiners who were immersed in CRM brought a lot more credibility. I was enjoying the simulator and the flying was varied. I was flying Gatwick-Sanford, which was our new airport in Orlando, and then Sanford-Manchester, and I was still doing all the usual European short haul. It was a lovely mix of flying and simulator.

In 2004, Britannia started Thomsonfly, a low-fares airline. It operated B737-800s, 500s and 300s for a few years. The B737-800 became the main short and medium-haul aircraft of Thomson Airways. Now Thomsonfly was being brought into the main airline. Some pilots had left Britannia to go to Thomsonfly for a quick command or the opportunity to become training captains.

The new chief training captain suggested we fly together to get to know one another. I could feel he was after something but I didn't know what. On the way down to Spain, he asked: 'Is there anything you particularly think needs sorting within training? Any ideas?' I replied that I thought we could do with the command course being changed a bit – and he said: 'Why don't you go and re-write it?' I replied that I wanted to look at some other airlines' command courses to get some ideas, so I went with colleague Mark Flynn.

I contacted JetBlue in the US, a fairly new and very successful low-cost airline with quite a buzz about it. They were based in New York but their training centre was in Orlando, Florida, and they invited Mark and me over. I wanted to see a two-crew dynamic and modern operation. They were flying Airbuses but that didn't make any real difference to a command course, as long as they operated jets. When we arrived, we were met by a chap who suggested we sat in on the morning part of the induction day. There were baggage handlers, office staff, pilots, check-in staff, cabin crew, the list went on and on. I thought: 'This is really buzzy, this is America at its best!' At the end of the day, they invited us to come back tomorrow to look at simulator training. Mark was thinking that would be nice, but I said: 'Sorry, we are committed elsewhere.' We said goodbye and Mark said: 'What are we doing tomorrow?' I replied that we were going float plane flying.

I knew Mark had lots of experience on float planes and I had flown a sea plane years before on Loch Lomond. Jim Beaton had somehow got hold of a Lake Buccaneer Amphibian, a four-seater with a high 'pusher' prop, high up behind the cabin. It was like a boat keel with retractable wheels. The trouble with a sea plane is that once you start the engine, you are off. Jim and I went down low-level past Ayr harbour and then pulled up, calling Prestwick Approach: 'We are a Buccaneer inbound to you low level.' They thought we were flying a Buccaneer military jet! We grinned like a couple of kids. You had to make sure when you landed on a runway that the wheels were down, and that when you landed on water the wheels were up.

Orlando is full of lakes, hundreds of them, and on one of the lakes there is a famous place called Jack Brown's. They had a couple of Piper Cubs, tiny little high-wing canvas aircraft with tandem seats on floats, and we said we'd like one each. Of course, we needed an instructor and they brought out these two children – they were 16! They jumped in the front and Mark and I climbed in the back of our aircraft. We got airborne, with me following through on the controls. I watched the air speed indicator just to get an idea of the speed which would let this wonderful little aeroplane fly. It wasn't fast, but with that information I knew I wasn't going to die if the lad in the front passed away. It wasn't too difficult in the benign conditions and away we went. We did a bit of formation flying and went over 'Fantasy of Flight', an airfield south of Orlando where they have a B-17 WW2 bomber you can go up in. They have a couple of big hangars which are open to the public and lots of aeroplanes, including a Sandringham flying boat. We decided: 'We'll go there next.'

I presented my report to the two chief training captains. The main points were that we needed more coaching and a bit more time for pilots to settle into the left-hand seat as captain. The chief training captain asked if I would be interested in chairing the command selection board. That role had always been carried out by the chief training captain within management. I had never been on the board. Now we were into the 'white gloves' area. The command selection board had six members who were all very experienced and well regarded TREs. I was a fairly new examiner, and he wanted me to chair the board. With that, I became chairman of the command selection board in February 2006.

Chairman of the board

The new role of chairman of the CSB (command selection board) made it necessary for me to be in the office at Luton much more and I spent a lot of time with the chief training captain, who had just become HOT (head of training). Again, he said: 'We'd really like you to be chief training captain B757/767.' I left for home thinking: 'If I don't do it, someone with no vision, no passion and no leadership skills might get the job.' I wasn't even sure I had these skills myself. I spoke to Ruth and the next day I went back to the chief training captain and said I'd thought about it but there were three conditions: I'd only do it for a year, I'd never undertake the role of flight duty manager... and the third I can't remember! The new HOT agreed to my three requests and I took the job.

The airline was expanding and the training department was busy, especially on the B737 fleet. We were partly supporting some trainee pilots through their professional pilot's licence. They were at a flying school at Jerez in Spain. They might start the course with no flying experience, and 18 months later leave with a frozen ATPL/IR, qualified enough to fly one of our jets after completing a type-specific training course which would last a further three months. They would then join the line as a new second officer.

The training department I had known when I was running ICT had gone. We no longer owned and operated our own simulator, so the important role of simulator manager had disappeared. By late afternoon on my first day, it became clear that I was the B767/757 training department. When I went downstairs to the section of the rostering department that dealt with pilot training, I asked: 'How many training captains do I have?' No one knew. It took days to get an up-to-date list of training captains, and it was immediately clear we didn't have enough. We were using retired Britannia type-rated examiners (TREs) in the simulator, which was much better than using just any old third-party trainers, but they were not flying the line, they were not current. Importantly, they were taking TRE positions that should have gone to our type-rated instructors (TRIs). The department was a shambles.

I could have done with a handover brief, but I hadn't heard a word from anyone. It was just: 'There's your office – carry on.' I thought: 'This can't be the way to do it.' It was clear we didn't have enough trainers on the fleet, so I started the process of recruiting TREs and line trainers. I hadn't been given any budgetary control. I didn't even ask for it. I just insisted that we needed more training captains. I said to my boss: 'You know this is in meltdown and there's no way it's getting better. I'm going to rebuild the department and it starts with a deputy and then a management team.'

The B737 chief training captain and I flew out to Jerez to pin wings on our cadets who were graduating, and there was a group of about 15 youngsters. The cadets were there with their parents and I gave a little speech, telling them how lucky they were to be joining us. I told a couple of wee stories and made them laugh and said: 'You are going to have to work hard, do a good job and look after the jet and you're going to have a wonderful career.' It was straight out of any company handbook, but if you do it with some passion, which I did, it means a lot more. I took them all out after the speeches and got them well oiled.

The HOT had asked me a couple of times if I would take over the B737 fleet as well, so I became flight crew training manager in September 2006. At last, I had my department and it was a miracle. It was not costing the company a fortune because all the managers except me were meant to fly the line as well. We began interviewing for training captains.

When I took over the B737 fleet, I found out within a week that some of the captains on the fleet had been cleared to operate into Madeira in a simulator not approved for that role. As Madeira was a Cat C aerodrome, the pilots needed extra training and, in this case, that included departures and arrivals in a simulator with an approved visual model of the airport and the surrounding terrain in both day and night. We produced a detailed training plan for Madeira. All further Madeira training would be done in a certified simulator. After we had sorted that out, I had the requirements for all Cat C airports detailed in the same format in the operation manual.

During my first year as flight crew training manager, we had an incident which was put down to poor pilot monitoring. Both the pilot flying and pilot monitoring had failed to notice the aircraft diverging from the correct flight path. How should we train and evaluate monitoring? We found a new device called Eye Tracker which lets you monitor on a screen where someone is actually looking. It worked by getting the pilot to put on a baseball-style hat with all sorts of gizmos on it which track his eye movements. It was essential that it was accurate enough to identify exactly what part of any particular gauge they were looking at. We did some testing, including one in the simulator on a randomly picked first officer who was operating as pilot monitoring. On closer examination, he turned out to be 'the pilot not monitoring'. He was working his way through the QRH (emergency quick reference handbook) checklist because they were dealing with multiple hydraulic failures. There were various lights on and failure messages, and they were dealing with complex procedures. It became clear that the pilot monitoring was completely wrapped up in the emergency checklist.

The proper way to use the QRH is for the pilot monitoring to place it in front of themselves and slightly over towards the other pilot. This way, you can illuminate the checklist and the other pilot can look over and follow the drills progress. At the same time, it's important to monitor the instruments and the flight path of the aircraft. It is a case of reading and actioning a bit, then monitoring for a bit, and then checking: 'Is the aeroplane still flying? Is it all okay?' If the answers are yes, then you can read a bit more.

If someone said to me 'You are on the approach and I want you to read something and not look up for 30 seconds', I would reply: 'I can't do it. I need to look up now.' My conscious thought processes would be screaming at me to look up. The trainer could see on the Eye Tracker that this lad was just reading and not scanning the flight instruments. He said discreetly to the captain: 'I'm going to fail the auto throttle. The throttles are at idle power but don't say anything.' The pilot monitoring didn't notice the speed loss or the high nose-up attitude. The pilot flying was told to ignore the further failure and about 50 seconds later the stick shaker activated - the aircraft was stalling. That is when the other pilot looked up from the QRH with some alarm. We hadn't known before that pilots were not monitoring to this degree.

This simulator session was set up to see how the pilot flying scans the flight instruments and not particularly what the pilot monitoring was scanning. I have been in the simulator training new pilots' take-offs and landings just before they go base training on the actual jet. A couple of times I've found myself leaning forward during the latter part of the visual approach trying to see where they were actually looking. If the approach was unstable, it usually meant they were flying the jet like a Cherokee, just looking out of the window! That's not how to fly a jet. You fly the jet by numbers: if there is an instrument landing system, follow it just like when you are flying an instrument approach. Your scan moves constantly from the flight instruments to looking out at the runway and back to the instruments. As you get closer to the ground, the scan is mostly outside with a wee look inside until the flare, which is done visually.

Eye Tracker showed that these airline pilots, who you would have expected to know how to monitor, were not doing so. We recorded and filmed the instrument scanning of very good pilots and found they didn't actually scan in the previously established method of the 'T' scan, which was looking at the ADI (attitude direction indicator), then ASI (airspeed), then ADI, then Heading, then ADI, then Altitude, etc. This was the scan we had all been taught during initial flight training. It wasn't wrong, but the good pilots tended to just look around in their own way, picking up the information they needed. Possibly, the new flat screens with all their extra information had developed new scanning skills. We could now show what 'good' looked like - and, of course, what 'bad' looked like. If the pilot flying's scanning is poor, the outcome is obvious. The speed control is poor, or heading or height control is poor. But if the pilot monitoring is not monitoring the flight instruments and the pilot flying is doing a good job of flying the jet, often through the autopilot, any poor monitoring by the pilot monitoring is completely covered up.

The good pilots, when actioning the QRH, might read 'Left hydraulic pump light should be on' and they'd look up and say: 'It is.' Then, before reading the next line, they'd have a look at the flight instruments and check all was well. They would read the next line, action that, and then take another look. We had no idea we had to train this and I never heard another airline say they had a problem, but this poor monitoring had been going on for years.

A little light relief

In 2004 I had been checked out again on the company light aircraft and I was delighted to be cleared to fly. Britannia still had a flying club and a couple of the aircraft were based at Luton - a PA-28 Cherokee and a Mooney. I started doing lots of light aircraft flying again, which I enjoyed immensely. I was based at Luton so it was convenient to park by the crew room and someone from Servisair, our handling agents, would drive me over to our hangar where the light aircraft were kept. I used to take my girls, especially Leigh, who has done a lot of flying with me over the years. None of my daughters has ever seriously expressed an interest in getting a flying licence and I haven't encouraged them. If they wanted to do it, of course, I would help them all I could, but I've spent my life coming across dozens of people who have said: 'I'd like to do that.' I've spent a lot of time with them, giving them tips and hints, and in the end none of them has gone very far.

One day, I dropped my daughter Ceinwen off at Filton. It had started off with clear blue skies and for most of the return flight it was gin-clear and the flight was smooth. As I passed Gloucester airport, the cloud was thickening and by the time I passed Turweston, I was on top of solid and rising cloud. I had lots of fuel and knew I could just turn around and divert back to Gloucester, or even back to Filton, but I thought Cranfield was worth a look. As I passed over Milton Keynes, there was a large hole in the cloud and I thought I would spiral down and sce if I could fly the rest of the way low level. Failing that, there was room enough to climb back up and keep clear of the ice-laden clouds all around me. I was just north of Milton Keynes and snuck under the clouds and landed at Cranfield. Good job! I had pushed the flight as hard as I could but I always had a good plan B. They've closed

Filton now and are redeveloping the huge runway that was built for the Bristol Brabazon. They demolished the village of Charlton in the late 1940s to extend the runway to accommodate the take-offs of this giant plane. The other Bristol airport, the international airport, is built on a hill and the runway is not overly long. It is also miles from the motorways. I sometimes wonder if they closed the wrong one.

I spent a lot of time at Turweston when I was a manager. When sitting in my office in Luton was all getting too much, I would take the company mobile and spend the late afternoon at Turweston. It kept me sane. It was in late 2006 that a group of us decided to buy our own light aircraft. Rob Burgess, who was a B737 training captain and kept his own aerobatic aircraft at Turweston, was wandering around one of the hangars with a mate one day. They were talking about putting a group together, based around one of the new Cirrus aircraft. The Cirrus was a whole new world compared to something like a Cherokee. It was fast, with four seats, a 200hp engine, a three-blade constant-speed propeller, and was controlled by a single throttle. It is built using a composite construction and has advanced aerodynamics. It is also fitted with CAPS, a ballistic recovery system. A solid-fuel rocket is used to pull a parachute from its housing and deploy the canopy. The whole aircraft then descends under the parachute – so final touchdown is like dropping it from 15 feet.

Word of the possible new group spread around the airfield and there was a fair amount of interest. There were two engines available, the larger 310hp and the 200hp. We didn't want the big engine, as the fuel burn would have put the hourly flying cost out of reach, but the 200hp version seemed just fine. I felt that, if I was going to be involved, I would want an autopilot fitted because whizzing around the London area single crew would be hard work. Uncontrolled airspace was getting tighter and I didn't want to infringe the controlled air space around Luton and Stansted when flying down to the south east coast.

We put a group together and I got my autopilot. In fact, we opted for the top-of-the-range Cirrus SR20 GTS, which meant it was fully spec'd. It had all leather interior and twin GPS, and even TCAS-like traffic information. Twelve of us formed a group to buy the aeroplane. Nigel, my brother who flew for BA, joined the group. The Cirrus was one of the new generation of sophisticated light single-engine aircraft. The CAA was introducing a requirement for further training and our insurance brokers also wanted some form of training for the pilots in our group. I spoke to the CAA and they agreed that I could check the group members out on the single throttle which controlled both propeller pitch and power. The navigation package and engine monitoring system were all presented with the flight instruments on two flat screens. It was quite sophisticated and the aircraft was quick, so if you were getting into difficulties, you were covering a lot of ground while you were struggling.

The aircraft wasn't terribly demanding to fly but its landing speeds were a good 15mph faster than the Cherokee. If you didn't have good speed control, your landing run could be long and/or the touchdown unstable, with the nose wheel possibly hitting the ground a bit hard with very expensive consequences. It didn't have retractable undercarriage. It had very tightly fared wheels and that's why it was not so good on grass. The Americans would have said: 'What the hell are you taking it on to grass for?'

New aircraft used to be ferried over from the States, but there was always some risk in flying a single-engined plane over the Atlantic. Ours, therefore, was one of the first to be partly dismantled after its initial test flight in America, containerised and shipped over to the Isle of Wight. There they were rebuilt, flight-tested and prepped for delivery anywhere in Europe.

Never easy teaching the boss

During 2007, while I was flight crew training manager, Thomson announced that it was merging with First Choice Airways. Throughout the merger, the CAA had to make sure all operational rules and SOPs (standard operating procedures) were understood and followed by all the flight crew. The two airlines were similar but the SOPs were quite different. The question soon arose: 'Do we operate to the Thomson SOPs or to the First Choice SOPs, or do we start again from scratch?' We couldn't shut down the airline while the pilots all learned a completely new set of SOPs, so the third option was quickly dropped. We were the bigger airline and it quickly became clear that we had the bigger and more robust training department, so it was decided to use the Thomson manuals.

The training department had to take two different airlines, with all their different standard operating procedures and cultures, and come up with a plan. It was the first time I realised that most of the type rating examiners were just normal fallible pilots, warts and all. I had always revered most TREs, but when I looked closely with a critical eye, there weren't that many that stood out as excellent and an obvious choice to join my management team.

When I was checked as a revalidation examiner TRE (RETRE), the training inspector (TI) was watching me, I was watching the TRE and he was checking the two poor old pilots at the front of the sim. The pilots under check hated it when all this 'examiners checking examiners stuff' was going on in the back. They felt that if they did the least thing wrong, they would fail. It was not really the case but I suppose if the TRE had any doubt, he would choose the fail.

By 2008, our merger was in full swing. I was now a RETRE and a

RE-CRMIE (revalidation examiner of crew resource management instructor examiner). In other words, I had all the tickets, all the badges. And I was very pleased. I couldn't help remembering my schooldays and I said to myself: 'Thank you, headmaster, for those encouraging words - bastard!'

I became manager flight crew training for all fleets in May 2008. I enjoyed the day-to-day bit and I enjoyed my team. We looked after the trainers and we were always there for them. By 2008, all the pilots were flying together, the 'pink' from the First Choice side and the 'blue' from Thomson - because our aircraft were blue and theirs were pink. The two airlines were well respected and both were stronger together. In late 2008, I thought: 'I'm going to fly the B737 because some of the trainers on that fleet need guidance on using good judgment when it comes to training and examining.' The CAA TRE course was all about failing the pilots - nothing is mentioned about passing. When they started Thomsonfly and introduced the B737-800 fleet, they had to bring in direct-entry captains, first officers, trainers and managers, and they just didn't have or respect our culture. As a result, the new guys were getting pushed on to the 'standards' courses and the outcome was sometimes a bit brutal. The Authority wanted us to use the terms of 'retest', 'fail', 'repeat', so my job was also to introduce that to the B757/B767 guys. There had to be some change on both the fleets. Another reason I wanted to fly the B737-800 was because I felt the trainers on that fleet were a bit ignored.

I decided to do the simulator training in Palma. I wanted the B737 chief training captain to do my actual sim training to see how he performed in the simulator while training me. I asked him to bring out one of the new TRIs so I could look at him as well. It was a great opportunity for me to have 'one on one' training and also to watch how they trained me. I said to my 'trainer': 'Now, son, I don't do early, so you may need to brief me the night before, depending on the sim times. We can have breakfast before or after the sim and then debrief either after breakfast or with coffee, in a quiet area around the pool.' I was immediately, and quite deliberately, putting them under some pressure. It's never easy to teach your boss. So off we went and I enjoyed it, although I was just a bit nervous.

The old B737 chief training captain told me: 'The B737 is much harder to fly, especially on one engine, and also when flying with manual reversion.' My immediate reaction was: 'Gee, thanks!' It is demanding, partially because it has only two hydraulic systems and if they both fail, you literally have rods and wires to move the flight controls. That means there is a big dead space when operating the control column in both pitch and roll, so the flying controls feel very sloppy. The B757 and B767 have three hydraulic systems, plus an RAT (ram air turbine) which drops down under the fuselage and is a small propeller driven by the air flow. The RAT powers all the flight controls if all three systems fail as there is no manual reversion.

The first day went well. I had put a lot of work into the preparation, and at the end of the day I was debriefed on a nice quiet sun terrace at the hotel, as I requested. When we got to the bit about the flight controls, the trick in the flight crew training manual - written by Boeing – was that you trimmed the aircraft out of the dead area. This means you either have to push forward or pull back all the time. I tended to use the forward one, so that if I lost concentration for any reason, the aircraft would pitch up rather than down.

My co-pilot wasn't being trained. He had been sent out to fly with me to fill a seat. I told him to fly my speed using the thrust levers. He would be my auto throttle and I would concentrate on flying the height and heading. The guy teaching said: 'Oh, no. We don't do it like that.' What do you mean? He replied that I needed to do it all, so I told him to stop the sim and tell me why. 'That's how we do it.' I said: 'Well, not any longer. It's a two-crew aircraft. I do the flying bit and he does the speed control.' He repeated that we don't do it like that, and I said: 'We do now, son!' 'Oh,' he said.

During the debriefing I explained clearly why the way I did it was better, and why there was no reason to make one pilot do everything while the other dozed in the corner. The next day, we practised emergency descents and the trainer said again: 'We don't do it like that.' I said once again: 'Stop the sim!' I didn't mind that he was teaching me the way he had been taught, but I did mind that when I told him a better and easier way, which Boeing suggested, all I got back was a protest. I wanted good, flexible training, with the trainers showing good judgment.

We were due to be out there for three weeks, with 12 simulator details. We did three days in the sim and then had a day or two off. There was also a bit of studying to do. For my final test, my boss flew out to do it and it was fine. After some line training sectors, we went to Faro for my final line check. I found the B737-800 to be a slippery wee aircraft. We had to use quite a lot of speed brake to slow the thing down. We landed, and I had to do a pilot monitoring sector to be signed off. All my sectors up to then were as pilot flying, so I had never raised or lowered the gear. We got airborne and it was a beautiful day. We turned out of Faro and I put the gear up. In the Boeing, the gear is hydraulically retracted and once it is up, it is held up hydraulically until the checklist requires you to position the gear lever to 'off'. Then the hydraulic pressure is removed and the gear remains held up by locks. Because the gear lever is more sensitive than the B757's, I missed the 'off gate' and selected the gear back down - so the wheel went back down, with a great roar. I let it cycle, checked the maximum retraction speed, selected the gear back up, and said to the boys: 'I'm going to do a PA and I bet you're going to both listen.'

I said: 'Ladies and gentlemen, this is the captain. If you look, you can see a gorgeous day. We're only climbing through 10,000 feet and you can almost see New York on the left and you can see Italy on the right. And, oh, you might have wondered what that noise was. It was just the wheels going down, but they're back up. And it's nice and quiet now!'

He signed me off, but I only flew the B737 for a few months. I tended to fly with training captains to get across the different culture and just chat and get to know them. I wasn't there long enough to know if I succeeded.

The B737-800 had a new wing, so we could fly as fast as a B767. There were new engines, a new flight deck with all flat glass panels, and a much better range than the old B737-200. It was almost a different aircraft. It would get to Tenerife with no problem - and it would even get to Sharm El Sheikh from Glasgow, with 189 passengers instead of the 130 on the 200. On the B757, we would burn a couple of tonnes more on a long flight than the B737, so it almost sipped fuel. It was quite unbelievable. However, it didn't stop as well as the B757. The B757 remains a hugely capable aircraft but it is heavy and you usually don't require its superior range or take-off and landing performance. It is far more economical to operate a B737-800 around Europe than a B757, but for destinations such as Cape Verde, or a bit further afield, the B737 runs out of puff and that's when the extra range of the B757 counts.

During my time in management, my boss would always talk about the needs of the business. Particularly after the merger with First Choice, he could see much clearer than me where the department was heading. Whatever they told him to do, he would try to do. My approach was somewhat different. It wasn't about wasting money but I just couldn't get away from trying to keep the old Britannia Airways 'gold standard'. I really struggled with second best just to keep the cost down. We were seen as a high-class department and I didn't want to change that. I was struggling with the constant pressure to reduce costs, course lengths and simulator time, and to use smoke and mirrors to make average look good.

The airline's fuel bill is colossal and even a tiny percentage of saving makes a big difference. There was a department whose task was to monitor fuel burn and come up with ways of reducing the amount of fuel used. Unfortunately, many of their ideas increased the pilot's workload. We were asked to use less flap on landing, to reduce the required engine power on the approach and therefore the amount of fuel burnt. This procedure makes the aeroplane less stable. Yes, the fuel bill was huge. Yes, it saved money. But the jet did require fuel and you cannot keep cutting. We were landing with a lot less fuel than we used to. Our fuel comfort blanket shrank over my last 10 years and that was okay - we always carried the CAA minimum, and often more if the pilots thought it necessary. But in the old days a captain would say to me: 'Flying the B767, you land with six tonnes, and with the B757, five tonnes.' We got used to not carrying 'a tonne for mum'. We thought about the amount and carried what we needed. So the company succeeded in reducing the average fuel carried and that was good - but there is a line in the sand.

Behind the jet

After three years in training management, I decided to go back on the line. I finished on the first of May 2009 and I was lucky to be able to go back to Birmingham as my base and fly the B757 and B767, though I remained in the training department and kept my role as a senior examiner, formally called RETRE. By 2012, EASA was becoming the European CAA and it wanted the three-yearly TRI and TRE check done separately. Up to that point, I could reauthorise a TRE's TRE and TRI on the same day. The rules were changing and the UK CAA was behind the curve. We had been told by the CAA that the check had to be over the two-day simulator check, so I carried out a couple over the two days. At the end of the second, I got a call from one of the CAA training inspectors to say what we were doing was not compliant. I said: 'You must be joking - and best tell the management.' In the end, the CAA said a senior examiner could extend the validity of a TRI's rating by three years without any test at all!

During that period, I had time to ponder how, before CRM had come along, the expression used was 'airmanship'. The problem with airmanship was that it meant different things to different pilots. CRM really defined airmanship. It broke it down so it could be understood in its various components. Our 'pilot skills list' took the CAA's version of the many CRM skills and clarified what we, as an airline, thought were the key skill areas. Then it identified what was required to achieve each headline key skill. This showed the pilots what 'good' looked like and could be used in debriefing to help develop their skills.

A trainer could, for example, identify 'situational awareness' (SA) as a weakness and, by referring to that part of the skills list, try to see where the problem lay. The headings under SA were 'Thorough pre-flight preparation', 'Stays ahead and updates plans', 'Makes contingency plans' and 'Keeps broad perspective'. Before the introduction of CRM, the trainer would likely have said: 'Your airmanship is poor and you are behind the jet. You need to be in front of the jet.' In this instance, the pilot would not be getting much help. There would be no tips on how to improve. It was a bit like saying to someone: 'You are ugly. You need to be less ugly.'

Situational awareness is one of the pilot's key skills and thorough pre-flight preparation is vital, but that will not cover all the various issues of dynamic flight. It took me about a year to catch up with the jet when I first flew the B737-200. The move from Loganair, where the aircraft flew at two nautical miles a minute, to the jet, which covered eight miles a minute, was a huge challenge. There wasn't a pilot skills list then but, looking back, I can see that slowly I learned the different skills now identified on the skills list. Some were easier than others. Some I was naturally good at and found easy, others were much more challenging. SA is vital to the safe operation of any aircraft – 'Where am I and where am I going?' These two questions should be running through any pilot's head at all times. 'Where is the high ground? Am I descending below MFA (minimum flight altitude)? If so, am I under positive radar control? And if so, where is the high ground anyway? Where is the airport in relation to my alternate airport and any high ground? Is my alternate going away from the high ground or do I have to fly over the high ground?' The list of questions goes on.

When I was the training manager, a fairly new line captain with a new co-pilot had flown into Sardinia. Sardinia has a lot of high terrain and obviously, being an island, there are areas of no terrain. They were required by ATC to break off their approach for some reason. They held over the island for a while and then radar vectored for an approach. During all this turning and orbiting, the first officer was asking where they were in relation to the runway. They landed and, as they cleared the runway, the first officer said: 'I thought we had landed on the other runway.' He had been confused and had lost all SA. The captain realised the first officer was 'behind the jet' but couldn't get the big picture across to him. One thing that might have helped was hand gestures. For example, 'the high ground is over to the left' and pointing clearly to the left while saying it. 'The sea is over there to our right' and, with large arm movements, clearly pointing to the direction of safety. I regularly used hand gestures in unfamiliar surroundings, even with experienced pilots. It helped me to build a clear picture in my mind of where we were, where we were going, and where the high ground was – and, more

Back in the left-hand seat.... taxiing out at Corfu airport in the B757

importantly, where it was not. The loss of SA is not uncommon. There is usually loss of SA in most crashes into terrain, but also a breakdown in many of the other skills from the pilot skills list. There have been many occasions when a captain has lost SA but the first officer hasn't. Sadly, there have been co-pilots aware of the high ground that was about to kill them but not daring to contradict the captain.

The pilot skills list also made quite clear that 'workload' was the key to success. In the simulator, I could identify the moment when the crew's workload stopped being managed, and the outcome five minutes later was a rushed and unstable approach. If the crew had made themselves time by going around the hold again and taking the time to brief properly and really think about the approach, their place on the 'arousal curve' would have been better. During the initial pilot CRM training, one of the skills taught was controlling stress. If you have no stress, you are usually dead but if you are extremely stressed and over-aroused, you stop functioning. Your ability to communicate fails, you can't speak clearly and you don't listen effectively, your heart rate increases and you start sweating.

At the bottom of the curve, you are dozing in an armchair or looking at the Milky Way on a quiet night flight mid-Atlantic. But as you near TOD (top of descent), you start to think about the descent and approach and briefing the other pilot, and you begin to move up the arousal curve. In normal operation, it is quite appropriate to move up and down the curve, depending on what is happening around you, but it has to be managed. If you haven't thought about it and suddenly you are turning on to final approach, the jet is getting away from you and you are likely to be high on the curve. If you have started the approach too far down the arousal curve and are not alert enough for the landing, then realise the jet is getting away from you, you move straight up the curve into a high level of stress. Then, on top of everything else, a warning message comes up. Now you will be moving into a dangerous level of stress where you cannot function effectively.

An example might be that you are operating a ferry flight, with no passengers or crew - just the two flight crew. The weather might be lovely, the aircraft fully serviceable and all is well with the world. You are flying from Manchester to Luton and you are pushing the aircraft, which is very light, with a very high rate of climb. You have left TOD to the last possible minute and require a very high forward speed to get the height off. You are enjoying yourself and working hard.

Then ATC inform you that Luton has changed the runway and you are now landing on the easterly. You have just lost 30 track miles to touchdown. Now you are seriously hot and high. Immediately, you pull full speed brake. Your mind is trying to do the maths - three times the height, plus at least 12 miles to slow down to get the first stage of flap out. You call for the approach checklist. You are asked three times to respond to the call: 'Altimeters.' There is no question about where you are now on the arousal curve. Minutes earlier, you had a big smile on your face. All the situation really required was for you to manage your workload. All it would have taken was for you to say to the pilot monitoring: 'Tell ATC I require an extra 30 track miles.' The lesson was clear – manage the workload, manage the curve, and you will manage your stress level.

Around the world

The merger had brought a new opportunity for us. First Choice had been operating VIP 'around the world trips' for a few years. The two lead American companies specialising in this field were now owned by Thomson. They were charging about $70,000 per head for a 24-day trip. There were also some trips around Africa and South America. They operated in the winter because that was our quiet time, and there might be 20 VIP trips a year. When I was a manager, I couldn't bid to operate a VIP flight because I would be away for a month or more. When I returned to the line in 2009, I bid for one departing in late 2010. It was a three-week trip around South America and it was a good one but before the trip departed, I went down with a bad cold. I was disappointed but deep down I had really fancied an 'around the world trip', a complete circumnavigation of the globe. I bid the next year and got one departing in February 2012.

The flight usually originated and finished in the US, and two captains and a co-pilot were required for each trip. The lead captain had to have done a VIP trip before, either as first officer or second captain, and there would be six or seven cabin crew and a load master, who was usually a volunteer operations officer. They were responsible for the loading of the aircraft, the flight crew's paperwork, and diplomatic clearances. There would also be two engineers and two chefs. The trip was arranged by National Geographic so the guides were of the highest calibre. The National Geographic experts would brief the passengers during the flight about their next destination. All the passengers had been given iPads linked into the aircraft's public address system which could be used during the lectures. Donald Johanson, a paleoanthropologist who discovered the skeleton of a female who lived 3.2 million years ago, was to be one of the two guides. The other was Paula Swart, an expert on Asian culture and on the people who lived on Easter Island who carved the gigantic stone figures (Moai). There was also a photographer, Amy Toensing, who was a regular contributor to National Geographic. Over the years, pilots have reported picking up some amazing people, including NASA astronauts who then lectured on the way to a particular destination. Picking up and dropping off the lecturers was the normal procedure but, on my trip, the two experts were on board the whole time.

I was a bit concerned about jet lag. We were to have a clear day at each destination and two clear days in Lima and Jaipur. I hoped that if I was careful, and didn't drink much alcohol, I would be okay. The flight was originating in Orlando so the crew all positioned out a few days before with Virgin. The aircraft we were using had been in Sanford for a few days and engineers had cleared all the snags, save one - but we didn't know about the ongoing issue at that time. Our cabin manager had done the whole trip several times before. All the cabin crew had been trained in how to deliver silver service. On one side of the aircraft, the Thomson Airways logo was replaced with TCS & Starquest Expeditions. This particular aircraft was named 'Explorer'. Inside, it was luxury floor to ceiling, with seats similar to first class. You wouldn't have thought that this was the same B757 that was flying the line weeks before. We used the B757 on the VIP trips because it had the range, and more hold space than the B737. The B767 would have been too big for some of the airfields we would be using. I was looking forward to the whole thing and to meeting what were bound to be very interesting travellers on board.

For the VIP trips, all the pilots had to be current ETOPS for the long ocean crossings. The lead captain on my trip was an ex-First Choice chap, a line captain and junior to me. I was no longer in management but I was a trainer, and I could imagine him thinking: 'Oh dear, I'm flying with Scott Birrell. This is going to be difficult.' The company arranged a briefing in Luton for the pilots about three weeks before departure. We were briefed about current world trouble spots and what problems we might encounter. The biggest concern was Luxor and Cairo and the company would be in daily contact with the Foreign Office. If Egypt was deemed too dangerous, we would go to Jordan and the passengers would visit Petra instead of the pyramids at Giza. We were also briefed on protocol between crew and passengers. During flights, we stayed out of the cabin because the food service was complex. Then the experts would spend an hour or so briefing the passengers, followed by a talk

by the expedition leader. Passengers and expedition staff would stay in the best hotel available and our crew of about 15 in the next best hotel. Since we would often want to visit the same places as the passengers, we were asked to leave the guests alone and engage with them only if they approached us. Unfortunately, nothing was mentioned about accepting gratuities from the guests - and that came back to bite us. We were all given a nice zipped folder, with a list of the guests and where they were from, names and pictures of the expedition staff, and a full itinerary of the destinations, hotels and places the guests would visit.
The night before departure from Gatwick, all the aircrew met each other. For the whole trip, we carried only $30,000 in cash. At one time, they used to carry $100,000 on these trips but more was now on account. The cash was there for unseen problems but the chefs would always need it for supplies. It was kept in a safe in the flight deck, chained and padlocked under one of the jump seats.
On the first night at Gatwick, the lead captain and I went off to the terminal to sign for the cash. I offered to look after the money and all the receipts to take some of the work from him but he said he would look after it all. I knew one thing straight away: if I ever had the opportunity to do another VIP trip as lead captain, the first thing I would do is hand over the cash to the other captain and say: 'You deal with it for the whole trip.' There would be scores of receipts gathered during the month and all would need to be justified. We were positioned to Orlando and the lead captain went off to meet the expedition leader. I said to him: 'Are you sure you don't want me to run the ship's funds, the cash?' He said: 'No, I'll look after that.'
I knew it was part of our job as pilots to welcome the guests on board. We were expected to meet them at the door, looking smart in full uniform. It was likely that among these people were maybe US senators, important sophisticated people. Some of the passengers had operated their own private jets. In one instance, an executive jet arrived, parked next to the B757, and a box of cigars was brought up the steps for the VIP guest on board. I quickly learned this was an entirely different ball game.
The lead captain said he didn't find it easy to stand at the door and welcome the passengers but I was happy to take that duty for the rest of the trip. The three of us had discussed how we were to split up the flying. The other captain wanted to fly the leg into Easter Island but didn't mind after that. There were going to be three opportunities for a bit of fun and low flying. These were Easter Island, the Great Barrier Reef, and Mount Kilimanjaro, and all would be weather-dependent.

Our first officer was quite happy to remain in the right-hand seat throughout, and the other captain and I would alternate in the left-hand seat. The first officer would fly the Great Barrier Reef sector and I would take the jet into Tanzania. We had a plan!

The world at my feet

We all had a couple of days in Orlando, then went to see the jet which was going to be our home for four weeks. I had never had anything like that before, our very own jet – we could leave all our things in the flight deck as if we were at home. I did a little 'welcome on board' speech, saying there would be three pilots and telling the passengers about the crew and the fantastic adventure they were about to embark upon.

Most of them were in their 50s, 60s and 70s, and almost all were Americans. There were a couple of pilots among the passengers too - ex-US Navy pilots who fought in Korea. I flew the first sector down to Lima, Peru, where we had two clear days because the passengers were going to fly on by a local airline and then catch a train to Machu Picchu. The first officer and I planned our PNR (point of no return) for the next leg to Easter Island, the most isolated inhabited island in the world, 2,400 miles off the west coast of Chile. There is only one runway and if we didn't have a plan, we might arrive to find the runway blocked. We wanted to establish a PNR which gave us the last point we could turn back on one and also two engines. I wanted to know when I was committed to a landing on the island. If, for some reason, the runway was unavailable, I knew we would be landing on grass. Like all things in aviation, if you have at least part of a plan, the outcome is always better than having to deal with something sprung on you at the last minute.

When the passengers returned on the third day, we departed in the morning on the next leg over the Pacific Ocean. There was little cloud and we could see the island 50 miles out. We descended to 1,500 feet and slowed to 170 knots with some flap out, flew around the whole island and then orbited out to sea and made another run around, so the passengers on the other side could get a good look. It was quite spectacular, zooming around low level and looking down on the hundreds of huge stone Moai statues, most with their backs to the sea. The island was very green, mostly grass but almost no trees. It looked as if someone had scattered stones and boulders over most of it. There was one small beach and the rest of the shoreline was rocky, with a huge crater at one corner of the island. We were flying the jet like we would be flying a Cherokee or a Cessna.

It was never going to be easy, not being the lead pilot. I was determined to be a good and supportive second in command. The first officer was a bit concerned that the lead captain and I might clash a bit and by the time we had reached Lima, we worked out fairly quickly that we needed to look after each other. We agreed that if either of us was ever getting a bit gobby, we would signal to the other to back off because we were going to have to make this work.

On our clear day, we drove around the island and admired the many Moai statues. We found the only beach and swam in the Pacific Ocean. The next morning, we fuelled the jet up to full tanks, another first for me, and launched for Samoa. The first officer was pilot flying and I was in the jump seat. The flight was planned at over eight hours. If the winds changed and our fuel reserve was getting tight, we would stop for fuel at Tahiti, but the winds were very light and we made it direct with good reserves. As we reached our cruising level and headed west, there was just ocean and the odd atoll below us. The Pacific is immense. We flew over Pitcairn Island, which is inhabited by the descendants of the Bounty mutineers. I was used to crossing the Atlantic Ocean in three or four hours. To cross the Pacific, it took us three days. We didn't hear one transmission until we got close to Tahiti. We arrived into Samoa in the early evening and rain was just clearing. The flight from Lima to Easter Island had taken 5 hours 45, then another 8 hours 30 to Samoa, and we were not even across the Pacific because we still had another six hours to go until Cairns. The stop on Samoa was for crew rest and fuel. We just couldn't go any further.

To get there, we had crossed the International Date Line. There was a 24-hour difference as we crossed that line. I hadn't previously thought about what it was going to be like. They had changed the position of the Date Line fairly recently to put some more islands on to Australian time, but I hadn't thought about the local time change as well - so it was a 24-hour change minus a few hours for local time. When we landed in Samoa, we taxied in and the first thing we asked was: 'What is the date? What time is it?' I was coping pretty well with the constant time changes because we kept going west. I tended to wake up early. The flights departed at a reasonable time and were usually not too long. In fact, I went around the entire world without the tiniest problem, even when we eventually arrived back in Orlando.

Samoa was a wonderful place, one of the main Polynesian islands. On arrival, there was no security at all and it immediately brought to mind what airports must have been like in the 1950s. We just wandered through, got on a bus and they took us to the hotel. It was late and most of the passengers retired for the night. It always took the crew a few hours to clean and prepare the jet for the next departure. No one came on the jet at any time so the passengers could leave whatever they wanted on board. Once we had decided on a fuel load and checked the details of the next sector, especially the diplomatic overflight clearance numbers, we would go into the cabin and give the crew a hand. I usually ended up wiping and drying the fold-up seat tables, under the demanding eye of the cabin manager. Our jet was the only aircraft on the apron.

When we left, I was at the controls as we flew further west, heading for Cairns in Australia. We were having a small technical problem that we noticed on the ground in Orlando but which got worse on the sector into Samoa. The crew oxygen was leaking. The gauge was reading 900 psi. The passengers have overhead masks which appear in an emergency and are fed by oxygen generators. The oxygen is then mixed with air. This is not under pressure and the supply lasts for 12 minutes. It's a chemical combination and it can even taste a bit chemically. As pilots, we had full oxygen, stored below the flight deck in the front hold. When necessary, the pilots wear a full oxygen face mask and we had a gauge on one of our screens which usually read around 1200 to 1600psi. When we first got on board the jet in Orlando, this gauge was reading just over 1000psi and we mentioned it to the engineers. They said we had a wee bit of a leak and we would have a new oxygen bottle waiting for us in Cairns.

If I looked up the official oxygen requirements, it would show that for four people in the flight deck we would need 860psi. With three people, we would need something like 670, and the minimum for two was 480. So the pressure level was always above that. We could have had four in the flight deck, which most times we didn't. When we did the low-level run around Easter Island, the flight deck was full, but at 1,500 feet, who needs oxygen? There was no oxygen servicing available at Samoa so we decided to carry on to Cairns, where the new cylinder was waiting. I didn't like to think too much about the problem, since the oxygen cylinder might not be the source of the leak. Airborne and heading for Cairns, we were in a part of the planet none of us knew. It was all new. The flight time was around six hours and the weather was typical of the sort of stuff we would find in the South Atlantic, with isolated thunderstorms, and quite big ones at that. We were picking them up on the radar but because it was daylight we could see them and avoid them visually. Looking out of the flight deck windows, there was just a huge amount of ocean. After a few hours, the oxygen pressure was below 800 and I had limited the crew in the flight deck to three. It continued to drop and we were now leaking quite steadily. I said to the lead captain: 'You are going to have to move into the cabin. There is not enough oxygen if we have a depressurisation. Bring in some portable oxygen bottles.' I knew things would have to be serious before we were going to have to start sucking that stuff.

At that point, there was a faint brrr sound and all the crew oxygen disappeared. We were a third of the way to Cairns and still over the ocean but our position was not far from Fiji. I was aware that some pilots would carry on regardless, but I was not prepared to cross my fingers because if we had a depressurisation, we would be dead very quickly. We were at 37,000 feet and useful consciousness was measured in seconds. A few aircraft have depressurised and, for whatever reason, the pilots have not got on to oxygen. As a result, they quietly became unconscious and the aircraft carried on until it ran out of fuel. I asked the first officer to refer to one of our route guides to see if Fiji had oxygen replenishment available.

Before our adventure, I had spent a week making up an A4 sheet with a map for every sector we would be operating, simply because I had never been to that part of the world before. If I had been told that we were passing French Polynesia, I would have had no idea where we were in the Pacific. I made out these colour A4 sheets with the route and all the alternates written in, including what runways they had, the length of runway and what let-down aids were there. At the beginning of every sector, we placed one of my sheets on the middle console and used it to maintain our situational awareness. At one point during the cruise, a passenger asked me where we were. I brought out my sheet and explained our position and what was around us, and they said: 'We'd love to have a copy of that.' So we had photocopies of these dozen pages made up for them. That became more than a thousand copies, because everyone wanted one, not just the passengers - and they all took the pages home with them at the end of the trip.

We decided we were going into Fiji, and the first officer was about to advise ATC that we wanted to divert. I stopped him because I knew they would immediately ask: 'What's your ETA? What's the problem?' When that happened, we would have no time to do anything. I told him: 'We're going to Fiji but it doesn't matter when we land.' We were still over the maximum landing weight and if we landed overweight, the aircraft would need an 'over weight' landing inspection. Unlike in the B767, I couldn't dump fuel on the B757. I said: 'Let's not land it overweight. Once we get the aircraft down to 10,000 feet, we won't require the crew oxygen any more and we can then slow down and dirty up (increase the drag by running some flap out), increase the fuel burn and reduce the weight.' I didn't want to drop the gear at that stage because flying at even 170 knots with the gear down was very noisy, although the fuel burn would increase markedly due to the drag. Once we had our plan, I called ATC and said we'd like to divert to Fiji, and told them what the problem was. We were given descent down to 10,000 feet so the oxygen problem disappeared. We told the passengers that we were going to hold for about an hour to burn off fuel and then we were going to land in Fiji. The flight to Cairns should have been about six hours and we took 4 hours 35 just to get to Fiji.

We landed in Fiji, taxied in - and all hell broke loose. Here was a Thomson aircraft on the ground in Fiji on the other side of the world. I waited a little while before I told operations where we were, just to cut down further questions. Richard, the lead guide, was asking: 'What is happening? How long are we here for? If we are delayed, we are going to lose the Great Barrier Reef.' We parked on a jetty and the lead captain came in and asked: 'Can we go low level to Cairns?' I said no. He said: 'I've checked the MEL (minimum equipment list) and crew oxygen is not in it.' I knew that was true - then again, neither is the left wing! The MEL is a huge book listing all the different systems on the aircraft. It tells us what we must have and if there are any temporary alleviations to get us back to a base where repairs can be carried out. The MEL doesn't include trivia like a cup-holder on the flight deck or things which are obviously required. I said: 'This airplane is grounded and it's staying grounded until I un-ground it. Is that clear?' He said 'Yes, Scott' and left the flight deck. The first officer looked over from the right-hand seat and said: 'You're absolutely right. But what do we do now? We are on the far side of the world.'

The lead captain was dealing with the expedition leader, while I was trying to get some information from the engineers. I thought it was time to break the news to Luton so I called and said we were on the ground in Fiji with no crew oxygen. You could almost hear the panic.

I started to think this was going to take some time, so it was best to get the passengers off. The other captain agreed and told Richard: 'This aircraft isn't going anywhere soon. Best start looking for overnight accommodation.' They came back saying: 'We thought of adding Fiji to our list anyway. We'll go and find the best hotel and we'll do something with the passengers.' The engineers, meanwhile, said they couldn't find the leak. They had waggled every part without success so decided they would fill the oxygen up again, do some more waggling, and after an hour, if it was still full, they would give the all clear. They filled it up and it was holding pressure but they didn't know why. If I got the passengers on board again and airborne, having been told that it was fixed, and then it popped again, I would have to divert and that wouldn't look very professional. I told the company that I was going to take the aircraft up to its service ceiling of 42,000 feet for a test flight and see if it held pressure. The company said: 'Do what you want, Scott.' They just wanted it fixed.

We got airborne, flew for 2 hours 10 minutes with the first officer and me in the flight deck at 42,000 feet overhead Fiji. Of course, there is no traffic up there. By then, we were both knackered. The other captain and the engineers came with us and said: 'It's holding.' We called it a night. The first officer took over for the landing and we left a fully serviceable jet and went to a hotel to rest. We needed a good night's sleep. We arrived back at the jet late morning and found the oxygen pressure was the same as the night before. I met all the passengers as they boarded and, of course, every one of them asked: 'Captain, we've had a great time. Is it fixed?' I told them: 'We've conducted a test flight up to 42,000 feet. I'm happy to go. Are you happy?' Resounding calls of: 'Yeah!' That was my get-out. If it popped, I would come back again and say: 'We tried.'

All this meant we had lost a night in Cairns and therefore our clear day there. The flight time to Cairns was 4 hours 50, the pressure held and the engineers changed the oxygen bottle anyway. The weather in Cairns was dreadful, with very heavy tropical rain, so no great loss. The next morning it was much brighter, with just broken cloud. The lead captain was in the left-hand seat and the first officer was pilot flying. We called ATC and said we were a VIP flight and would like to do a low-level run over the Great Barrier Reef and atolls at 1,000 feet. It was spectacular.

We flew on to Cambodia, with a flight time of 7 hours 20 over my old stomping ground Indonesia, which I knew from my days flying on the Hajj. We left Australia over Darwin and headed over Ujung Pandang on the island of Sulawesi, then on to Vietnam and finally Cambodia. We had a clear day there and the first officer, one of our stewards and I went to the Angkor Wat temple, built in the 12th century. We hailed a tuktuk and he took us to the temple complex, which was absolutely fabulous.

Next day, we departed for Kunming in China, which turned out to be an industrial town with tatty high-rise flats right up to the airport fence. We landed there because most of the passengers were flying on with a local carrier to see the 'gorillas in the mist' or visit the Great Wall. As soon as we crossed into China, we had to convert our flight levels from feet to metres, so the three of us were checking each ATC clearance to confirm the clearance given in metres was what we were flying in feet. The lead captain flew it in there and did a good job. We had two nights and a clear day off, so we had a guide who was also 'looking after us'. We went for a three-hour drive to a place called 'the stone forest', which was formed when the place was under water. It was more interesting to look at than the places we were driving through - lots of ugly high buildings. I wasn't disappointed to leave that city. The atmosphere from start to finish was all a bit strange.

Luxor laid low

We left China to fly to Jaipur, India. I was in the left-hand seat and the first officer was pilot flying. Once we had departed Chinese airspace and reverted to flight levels in feet, there was a sigh of relief from all of us. Indian ATC can be challenging and their spoken English is sometimes difficult to understand. We all remained on the flight deck so we could listen in to the clearances. As we approached the point where we wanted to turn left towards Jaipur and away from Delhi, we couldn't get a word in. The frequency was crammed with Indian chatter. We said we required a turn and requested descent, and all I got back was: 'Stand by.' This went on for an age. The lead captain suggested a call on 121.5, the international distress frequency, and this time they came back fairly quickly. We descended into a hot and very hazy Jaipur and were told to park on the military apron after landing. The passengers were bussed off quickly but it took ages for all the paperwork to be completed to the local official's satisfaction. The flight was initially planned to go to Agra and the passengers were to go to the Taj Mahal, but the runway surface there was deemed unacceptable by the company. They had been re-laying the tarmac and done the work poorly - halfway down the runway there was a lip of concrete.

There were two clear days, which gave the passengers time to travel by road to see the Taj Mahal, as well as Jaipur. Some of the crew went on a day trip to the Taj Mahal but I decided not to go because I wanted Ruth to be with me on my first visit. The first officer had been to Jaipur before with his family and had used a local guide who had been incredibly helpful when their son was ill, so he contacted this same chap and he met us at the hotel the first night. He said he could organise whatever we wanted. He organised for some of the crew to go to the Taj Mahal and set up a coach for the next day for me and the rest of us to do the sights of Jaipur.

The next flight, out of Jaipur, was my turn at the controls. Departing Jaipur brought endless forms, procedures and rules to be followed. All pointless but, apparently, absolutely necessary. It took nearly two hours to tick all the boxes. The lead captain had done the low-level flight around Easter Island and the first officer had flown the low-level run up the Great Barrier Reef. Now it was my turn, weather permitting, to do something different. We were heading to Kilimanjaro, Tanzania, and if we could see the mountain, I was planning to give the passengers a stunning view of the summit. We flew across the Indian Ocean and over the coast of Somalia. We could see the peak of Kilimanjaro a couple of hundred miles away. Because it is so vast, it straddles two countries. We called Kenya and said: 'We are a special VIP flight and we would like to fly around the summit before landing at Kilimanjaro.' They replied, rather kindly: 'Do what you like!' We then called Kilimanjaro ATC and requested the same. They gave the same answer so we descended to 21,000 feet. The mountain is about 19,500 feet high and we were 1,500 feet above the summit, which was covered with snow. I was just flying round at 220 knots clean. I couldn't select any flap since I was above 20,000 feet, a limitation set out by Boeing. We spent about 20 minutes low level at 21,000 feet. The passengers were able to take unbelievable pictures, then we dropped down and landed into Kilimanjaro in Tanzania, where the temperature was well over 30C. The passengers were met by a fleet of single-engined aircraft and they went off into the bush.

Next morning, we went to see the famous Ngorongoro crater. As we reached the top, we could suddenly look down into this vast crater with 600-metre deep sides and 300 square kilometres in area, a vast, extinct volcano. It was explained to us that the animals living in this huge crater never left. Whether this was because of the steep sides, I don't know, but there they remained in their own isolated world. As we dropped down the side of the crater, within minutes we could see prides of lions lying by the track, resting and thinking about lunch - because lunch for them was all over the place. The crater was teeming with animals. We drove through it throughout the day and saw huge variations of wildlife in a vehicle with the roof lifted up so everyone could look out and take great pictures.

Next day, we flew to Luxor, Egypt, where we were going to spend two nights before moving on to Cairo. The stop at Cairo was only going

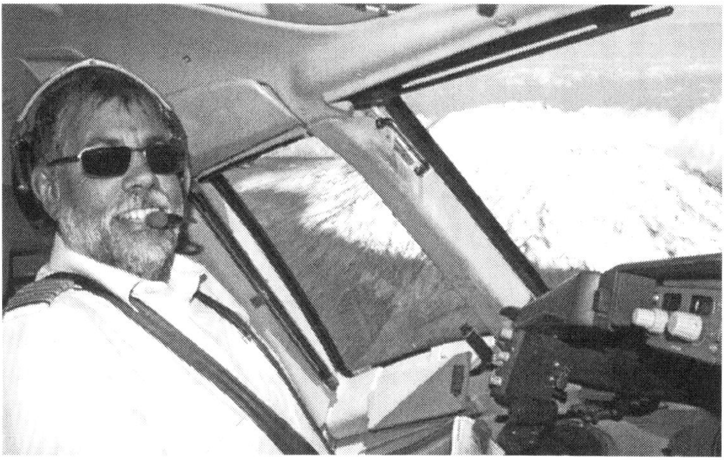

Orbiting at 1,500ft around the spectacular peak of Kilimanjaro

to be for a few hours and then we were due to fly on to Marrakech, Morocco. Ruth and I had done a Nile cruise five years earlier and when we visited Luxor, there were about 300 floating hotels operating on the Nile and the whole place was booming. On my return there during the world tour of 2012, the Arab revolt had kicked off and the number of boats there had dwindled to little more than 30. Worse than that, Luxor looked run down and empty and there weren't any tourists. In India, I had a tailor make me a black silk dishdasha Arabic robe. On our first night in Luxor, I headed off into town wearing it. I hadn't gone very far before a guy stopped with his horse and carriage and just looked at me. I said: 'Take me somewhere for a quiet beer.' In the carriage he told me the place was terribly depressed because the tourists had stopped coming. There was no money and people were getting really desperate. We stopped and had some beers and a pipe, then he took me to his uncle's shop and I bought an Egyptian keffiyeh head-dress, which made me look like a real local. I bumped into my crew later and they had a real laugh at my new look.

Next day, the carriage driver from the night before was due to take me and the first officer for a ride around Luxor, but he didn't turn up and someone else was there. We were taken around by this guy in his carriage for about an hour. Initially, he said: 'Anything you give us, even a dollar, will mean a bit of food for the horse.' He seemed really desperate. When we got almost back to the hotel, he stopped down a side street and I could sense something fishy was going on. I could see the hotel just up the road. He turned around and said: 'That will be

£100.' For a horse ride? I reckon I could have bought the whole district for less. We clambered off and he began trying to drive his horse in front of us to stop us walking away. We kept on marching towards the hotel. At one point, he grabbed the first officer's camera and was trying to pull it free. We just knew we had to keep going, which we did. In the end, we gave him a tenner and said: 'That's it!' We reached the main road, crossed quickly to the hotel and called for the guards.

We flew to Cairo and had three or four hours on the ground. The passengers were going for a quick whizz around the pyramids. When they left, we asked our handling agent to arrange a quick trip there for us. Six of us jumped into a minibus, had a very quick tour and got back to the airport before the passengers arrived back. There was the same feeling of poverty at the pyramids as in Luxor, and I was pleased to get back on the jet. We were going to fly along the North African coast to Morocco and into Marrakech - a 5 hours 40 minutes trip. The other captain was pilot flying on that leg and did a good job with the arrival into Marrakech. Moroccan ATC isn't the best in the world and they weren't having a good day, with confused approach clearances which served only to confuse other aircraft as well. Marrakech was another clear day. I hadn't been there before and it turned out to be lovely, a real French/Arabic city. We loved the souk, had a good look round, and enjoyed the best and cheapest tajine full of savoury lamb.

There are no aviation traffic rights between Morocco and the US, so we were going to have to stop somewhere which did have traffic rights. That suited us, because the B757 would have struggled to fly direct to Orlando. We stopped in Lajes in the Azores to refuel and all the passengers and crew had to clear customs so the flight would originate from Lajes, not Marrakech. Lajes is one of the regular nominated 'suitable on route alternates' which we require when crossing the Atlantic on ETOPS. I had never landed there so I was pleased to have the opportunity to see the place in daylight. The passengers were all off and back on and the aircraft refuelled and we were away in under an hour. They had it down to a fine art at the airport, so obviously this kind of thing happened regularly.

I was pilot flying for the last sector to Sanford, Orlando, from the Azores, which took 7 hours 25. Our route off the Atlantic took us just to the north of Cape Kennedy, so we called up and said: 'Can we do a lowish overfly of the launch site?' They replied: 'Sure!' So I flew the jet over the Saturn V launch pads where the astronauts had been launched to the moon, and on over the Shuttle's runway at a few thousand feet. We were then handed over to Sanford for a landing on the westerly runway. The next day, the other two pilots flew the jet over to Seattle on the west coast and because it wasn't a very long flight, I wasn't needed. The next VIP trip was due to start from there a few days later. The rest of us positioned back to the UK by Virgin the next evening.

The whole experience demonstrated to me that there is a very lucrative opportunity to establish a good niche market for round-the-world tours, where people were willing to pay $70,000 a ticket. Typical among our passengers were a middle-aged couple, both surgeons in the States who had worked very hard for years. They treated themselves to a world trip almost every year, picking different trips and routes each time. They wanted to see places most people didn't see. I thought afterwards that, as a concept, it would be quite possible to run a couple of aircraft all year round on these trips. We owned Starquest and TCS, but they didn't give us all the 'round the world' work. They sometimes used Iceland Air's B757s on the VIP flights. I was told by the chefs that the configuration of their B757s was better than ours because they had a central galley, so the cabin crew could arrange the meal service better than from a galley at each end. On the other hand, I believe the Americans would rather have a British crew.

The whole trip had been an amazing adventure and an unforgettable experience. I would have loved to operate another one before I hung up my flying gloves for good – though the next time as lead captain.

Accuracy is the watchword

In spring 2012, I had returned to Birmingham in my role as captain and examiner. The following year, I didn't do an awful lot of flying because I was the only senior examiner on the B757 and B767 fleet and in a position to sign Type Rating Instructor licences. In June and July 2013, during our busy period, I was just sitting in Gatwick signing licences. It wasn't arduous, sitting in the hotel bar drinking endless cups of coffee, but I wasn't doing enough flying. When I did fly the jet, I tended to say to the first officer: 'I'm going to fly it both ways.' They didn't mind but it was not the normal procedure. All the time at the controls I had to think: 'What do I do now?' It wasn't that I had forgotten how to fly the jet - that is like riding a bike. It was the many little things I had to do, like a PA to the passengers at top of climb which was the signal for the crew to bring us sustenance. I realised: 'I need to take control of this because I don't really want to feel slightly behind the jet for the rest of my career.'

Many years before, there had been a guy who resigned from the training department and was a very well thought-of trainer. He said he was coming to the end and just wanted to fly the jet. He also told me he had bought a caravan and just wanted to enjoy himself. I remember thinking: 'Wow, what a confident thing to do. I can do that.' I had to give six months' notice but my mind was made up – I was going to fly the line and I was going to enjoy myself. I bid to go on the B787, became a line captain on 50% part-time, and left the training behind.

I still got the odd phone call, mostly from trainers asking for a bit of advice, which I was happy to give. One captain phoned me about whether to extend his flight duty period by three hours. I told him that if I was ever not sure of something, I imagined being in a courtroom with a barrister cross-examining me about what I had done. I imagined the barrister saying to me: 'Captain, do you really feel you were rested after you had flown the whole of the North Atlantic overnight, then had to carry out a difficult approach in poor weather and then divert to your alternate, to then fly back to original destination? Surely a normal person would want to lie down?' Having said that, one important point to remember is that the only way to keep aviation 100% safe is for the aircraft to stay on the ground.

In the early years of the 2000s, I was a line training captain and asked to fly a B757 from Manchester to Amsterdam. It was a white tail aircraft without the usual markings, which meant the whole aircraft was white. The company had dry-leased it, which meant we crewed and operated it just like the rest of our fleet. I had been called out on standby and was told I was going to fly with another captain and would be in the right-hand seat. The captain was senior to me and because it wasn't a training flight, he would be the commander. I had no idea of the history of the aircraft, other than it had been in Manchester all summer and we were returning it at the end of the lease to leave it in Amsterdam. We did the take-off and as we were accelerating down the runway, I noticed that the air speed indicators (ASIs) were slightly erratic. Because the aircraft was light, the take-off speed was low and the acceleration was spectacular, and I didn't really have time to clearly identify the problem. As we rotated, my air speed indicator was reading that we were slowing, rather than accelerating, and it quickly became apparent that we had a problem.

I glanced at the left ASI and it was accelerating, but reading a different speed to the standby ASI between the two pilots. I leaned up over the captain's head to the IRSs (inertial reference system) on the top panel and selected ground speed. The ground speed, derived from the three laser gyros, was completely independent from the airspeed instruments. The IRSs couldn't give airspeed but ground speed would be a good gross error check. One minute, his reading would say 250 knots, the standby would read 220 and mine would say 150. And then, seconds later, it would be the other way round. I put out a Mayday call and Manchester ATC acknowledged. I said to the captain to forget the standard instrument departure (SID) and fly straight ahead. I then told air traffic we were climbing straight ahead. It was a lovely evening and for about an hour we flew around to the north of Manchester trying to work out what was going on. The ASIs would settle down, and then suddenly there would be a change and the captain was saying: 'I think the left one's correct.' I kept saying: 'None of them are right.' We asked to be radar vectored round for an ILS approach back into Manchester and radar said they could give us our ground speed down to about 100 feet, which was good enough. We were using the QRH 'Flight with Unreliable Airspeed' section to confirm pitch, attitude, power setting and flight path for our present weight. As they radar vectored us round, I was really concentrating on the IRS speed. We landed without drama and taxied in.

We discovered later that the aircraft had been sitting on the ground at Manchester for quite a long period before we had flown it and the problem was with the static vents. Airspeed in an aircraft is measured using multiple pitot-static systems. The pitot tube measures the ram air pressure and the static pressure is obtained through flush-mounted holes on the fuselage. These two pressures are fed to the air data computer and it is shown on the two main ASIs. The static pressure is also used for the altimeter. There are different failures, depending if it is a blocked pitot or a blocked static port. The blocked static port affects all pitot-static instruments and in the case of the ASI, as the aircraft climbs, the airspeed will read less than it actually is. There have been numerous accidents due to a failure in the pitot-static system. As we were in dynamic flight, we couldn't work out what the problem was. We have three systems and, in theory, and during all simulator training, we had only experienced a failure of one main system, leaving one system and the standby. Pilots do practise a loss of all airspeed but that is accompanied with loads of failure flags, so there is no doubt what the problem is. We had none. Since the weather was fine, we didn't really have to spend the time troubleshooting and thought the best thing was to get back on the ground. After landing, it was discovered that the static tubes had been blocked with ants, which had built nests while the aircraft was unused.

If an aircraft is to be parked for an extended period, especially in hot countries where there are lots of creepy-crawlies, the pitot-static ports would be covered. The problem wouldn't have been apparent during the walk-around unless someone spotted a line of ants high up on the side of the fuselage heading into a tiny wee hole. What I learnt from that was that, in any given circumstance, I had to understand where I could get the correct information I required. What this captain failed to do was accept that he had no air speed indicators. After that, I always took off with my IRSs, which are on the panel above my head, selected to show ground speed. I then had instant information without using multiple key strokes on the flight management computer.

Thinking about ground speed measurement always puts me in mind of my favourite watch, my Breitling Navitimer. When I did the round-the-world trip, I thought: 'I am going to need two watches – one for local time and one to remain in GMT.' So in Orlando, the day before we left, I bought a Timex watch in the 'President's Day sale' which was half-price at 20 bucks. That little thing has never lost a second since.

Complacency is the killer

In 1999, one of our B757s crashed at Girona, in Spain. It was a night flight, being flown by one of our well-respected training captains. Girona is quite challenging as Air Traffic is pretty average, there are high mountains in the vicinity, and the navigational aids are limited. There is an ILS at one end and it is one of those airfields you have to treat with a bit of respect. You come in quite high because of the mountains and you have to lose a lot of height, and then you shoot an ILS approach. On the night of the crash, very late in the approach during the flare, the airfield was struck by lightning and all the lights went out. It took quite a few seconds for the back-up power to kick in, and in that time the aircraft hit the runway hard, then veered off and on to the grass. It was just about in one piece but parts of the fuselage were at three or four different angles. People looking at this wreck must have thought there had been many deaths but thank goodness there were none.

In the analysis afterwards, they started to look at something called the 'startle factor', which we had never really heard about before. According to some experts, if someone receives a very severe fright, their mind almost freezes for a few seconds, and then they recover. There was a theory that this had happened to the captain, who was pilot flying. In that small amount of time, the aircraft hit the ground. When they started to wind back on events leading up to the crash, the question was asked: 'Should they have been on the approach?' Some time before this, I was in the descent into Girona and ahead of us was a huge thunderstorm. I thought: 'I'm out of here!' We diverted to Barcelona. I believe I was the only aircraft to divert. Wrecked or damaged aircraft are very rare indeed but there was another incident when an aircraft had a hard landing and was damaged. The question I asked myself when I considered both of these was: 'Was that approach within limits?' The answer was yes. 'Was it legal?' Yes. 'Would I have done it?' Possibly not. I stress the word 'possibly.' That's because the yellow streak on my back was getting bigger! It can be a very difficult call to know just how hard to push the flight and when to revert to plan B.

The storm at Girona might have looked benign according to the 'painting' on the weather radar, but although this was a fantastic aid to flying, every pilot knew the efficiency of the weather radar in each B757 varied from jet to jet. The hard landing again was not clear cut. The wind was strong but within limits, and the approach might well have been stable until a sneaky gust caught them by surprise. There are clearly examples when an approach should have been aborted but not necessarily in these two cases. I can't assess what the gut feelings of the pilots were as they made their approaches. The pilots at Girona couldn't remember what happened and the next thing they knew, they had hit the ground. Boeing was involved in the investigation and so was the CAA. The state of the airfield lighting was poor and so was Air Traffic, because they weren't giving out good weather information. After identifying the 'startle factor' problem, the question arose: 'How do you train for that?' The official finding wasn't pilot error.

As a first officer, I flew with all these captains, all different personalities with different capabilities. Some were excellent pilots, some were not. Back then, when I was a young lad, there were 500 or 600 pilots in the company and we didn't fly with the same people very often. There were two chaps I particularly remember flying with, both for sad reasons. We were coming back from Tenerife and the captain was talking about how he had been in Transport Command and he was going on about flying the Bristol Britannia, when in the RAF, all around the world. I was listening to these stories as we were trogging back at night from Tenerife, and I remember the sun going down and him talking about all these wonderful things I had never done.

Not long after that, he killed himself in a car, in a garage with a tube in the window. About a month later, another captain I had flown with did pretty much the same thing. They seemed absolutely fine to me. Obviously, they were not, but both were flying only days before taking their own lives. There was a lot of talk in the pilot community about why this had happened, but there wasn't anything amiss. It was just their private crisis. I was still young but the thought crossed my mind that at least they killed themselves without taking anyone else with them. It is recognised that most hull losses are due to pilot error. That must include the pilot's mental health, as well as poor decision-making, judgment, physical health, and aircraft handling, and, to a much lesser extent, poor technical knowledge. My gut feeling is that poor judgment is the biggest problem. All pilots these days have been trained in the various CRM skills and know that poor judgment may be the outcome of not properly using the CRM skills. All pilots can recognise the physical feelings of the aircraft running away from them. They can understand after the event why they shouldn't have shot or continued with the approach. The desire to continue and land is often described as 'press-on-itis' or 'get-you-home-itis'. There is also the element of being 'macho' – 'I will have failed if I throw the approach away and divert' or 'no one else has diverted'. They might feel they would lose face if they didn't get in or they weren't good enough, especially if other aircraft are landing. That's not the way to think of it.

Our airline had a clear rule that you must be stable on the approach at 500 feet - if not, you went around. The rule was there to stop unstable approaches leading to a possibly dangerous landing. It doesn't take a lot to make a well-planned and well-executed approach unstable. Maybe the wind changed direction and increased in velocity. There could be a billion different reasons. You can be struggling with a 20-knot crosswind one day and thinking: 'Why am I struggling?' It might be that the wind coming from a certain direction is rolling over the hangars and giving you problems at the flare, whereas a 40-knot wind might actually be better. It is not an exact science.

There is a big emphasis within the industry on pilot monitoring. Eye Tracker opened our eyes to the problem of poor monitoring. A Turkish B737 landed short of the runway in Amsterdam. There were three experienced pilots on the flight deck but through very poor monitoring of the aircraft's flight path, they crashed. It was due to pilot complacency. In aviation, complacency is the killer. That is the one issue which has to be tackled globally. All modern aircraft are well-built and incredibly reliable. Therefore complacency creeps in. It builds up over time and pilots need to be reminded that any aircraft will bite you, given a chance.

On top of these failings in pilots, some things are just damn stupid. We know the captain was in the crew bunk at the onset of the problem that brought down the A330 over the Atlantic. Crew bunks are often down the back of the aeroplane - the worst place. Where do you want your sleeping captain? Yes, he has to sleep, but somewhere near the front? In a ship, the captain's bunk is close to or on the bridge. Many aircraft have been lost because all the holes in the 'Swiss cheese' have lined up. All it would have taken is for one of the holes, one of the events, not to have happened and the aircraft wouldn't have been lost. Something fairly minor can save a jet, something the bean counters think is okay and legal to get rid of might just make a difference on a dark and dirty night.

Madeira is not a piece of cake

Some airports are easier than others to operate into and out of. All airports are categorised, and the three main categories are A, B and C. Newcastle would be a Cat A airport because it has precision approach facilities, good runways with no performance limited procedures, and also good night operational capability. Birmingham is Cat B because it has at least one of the following: unusual local weather, non-standard approach patterns, performance limited runways, or obstacles in the vicinity. Madeira is Cat C because it has clearly identifiable characteristics, needing specific training and briefing. The Portuguese have laid down the training requirement for a captain to be cleared to operate into Madeira. This includes simulator training and a flight into Madeira with a training captain. There is a requirement of at least one visit every six months, otherwise further simulator training is required. All take-offs and landings must be carried out by the captain.

The runway at Madeira runs along the side of a very steep hill, and about half of it is built on concrete columns over the water. Both ends require a curving approach and at each end there is a valley running down to the runway, which exacerbates the very tricky winds. I have landed there and immediately had to go around. As I touched down, I had a 10 to 15-knot headwind which instantly turned into 15-knot tail wind. I could immediately feel the aircraft was not settling on to the runway, so I went around. I only understood exactly what had happened because the safety department downloaded the data from the aircraft after I submitted a flight safety report and they could see the dramatic wind change. Going in on the north-easterly runway, you come in over the banana packing sheds at about 700 feet, where the approach lights start, strobe lights curving along the shore. You fly a curve in a bank to the right with the left wing high, covering the terrain that rises on the left. But you can't see it - and the runway, which is on stilts, is over at your two o'clock. There is a valley at your 10 o'clock, which the wind can funnel down and spill into the bay just as you roll the wings level. I used to look at the water and see swirls on the surface where the wind was almost spinning the sea. It was a good clue to the turbulence we were about to hit. We were also cleared to operate at night, so it was the sort of place for which you just had to have great respect. The other end was equally demanding for different reasons. You were running in and paralleling the cliff face with your wing tip. It must have looked very impressive for the passengers on the right-hand side.

Skiathos is also Cat C but doesn't require any simulator training, only an airfield visit with a training captain. Corfu isn't a Category C but it is a Category B, and for good reason. On one end it's not bad, but the runway is in a terrible state, in a lagoon so there is water on either side. We used to find the odd Greek jet sticking out of the lagoon. Coming in on the southerly runway, there is a hill and as you descend on the final approach, you are flying down the hill, which looks very off-putting. There is also a displaced threshold and the precision approach path indicator (PAPIS), a system of red and white lights that provide guidance to maintain the correct vertical approach angle. The lights are pretty dim and useless until you are close in. It is one of those places, especially at night, where you have to have a plan. You have to think about what your rate of descent (ROD) should be from base leg and all the way to touchdown. You need to take your time and always remember that if you don't like it at any moment, head off somewhere else.

The other problem is that the nearest airfield with decent approach aids is Brindisi in Italy and the nearest one in Greece is Thessaloniki. Thessaloniki has two good approaches but the approach on to the northerly runway is very demanding. You need to sit back and say: 'If I don't get into Corfu and, due to weather, I want a precision approach, can I fly an ILS into Thessaloniki or am I going to Brindisi?' There is no point leaving an airfield with poor navigational aids and diverting to another with poor aids. Unless the weather is fine, I have given the advice: 'If you are going somewhere where you've got a non-precision approach, make sure your alternate's got precision approach. If your destination runway is short, make sure your alternate has got a long runway. Otherwise, when you call for flaps one and nothing happens, you might be committed to a high-speed flaps up landing on to a short runway – and your day will have just got a whole lot worse.' At JFK airport, New York, there is a non-precision approach which we used to fly on the old B747-200. It was called the Canarsie approach on to runway 13L and 13R (two parallel runways left and right heading 130 deg) – a VOR approach followed by a circling approach. Minimum descent altitude (MDA) was at 90 degrees to the actual runway. The whole approach was a long curve. When you reached MDA, the runway was way off to the right and looked like a road in a built-up area. The approach is even more demanding with a southerly wind and usually not helped by ATC constantly telling you to keep your speed up. Some lunatic had designed it this way partly to keep the approach over the bay, thus keeping the noise down for the locals. Also, if JFK reverted to the precision approach for landing in this direction, the ILS and missed approach procedure infringed the operations at both La Guardia and Newark airports. So sometimes our workload was increased for local issues. If aircraft started missing the approach or pilots requested the ILS approach, it would be changed.

LAX at Los Angeles was the sort of place where if you went in there a couple of times a month for six months, you would have seen and done it all. But if you only went in occasionally, it could be a bit challenging. ATC once told me when taxiing in after landing: 'You're cleared to the 50 yard line.' I asked the captain where that was, and he replied: 'It's not on any map.' I immediately thought: 'They're having a laugh.' It was a typical American assumption that you were completely familiar with the place.

Microbursts can be shear hell

We don't get big thunderstorms in the UK, hence we don't really have to deal with microbursts. But you do in the States or in the lower latitudes, where the tropopause is higher and consequently the thunderstorms are bigger. Microbursts happen when huge amounts of air and possibly rain and hail come out of the bottom of a storm. The air hits the ground and bounces and curls back out and up. If you have a thunderstorm near an airfield, this descending air is likely to be at the core of the thunderstorm and heading straight down. Turbulence can and will affect an aircraft in flight in many ways. One of the most powerful elements we can experience is a downdraught, part of what is called a microburst. We practise how to handle and recover if caught in a microburst during simulator training.

As you approach a thunderstorm cell, the first burst of air movement is a rising air current which increases lift and your speed increases. You reduce power and push the nose down, trying to stay on the glide path. Then comes the horizontal air current, which will help regain your bug speed so all should be looking okay, except that your thrust levers are closed or very nearly closed. All of this is happening close to the ground and late on the approach because it is the descending air from the cell hitting the ground which is causing the effect. Of course, the next air you fly into, none of which you can see, will be the down draught. The aircraft loses lift and, unlike with a piston engine where the power is instantaneous, the jet engine can take seven seconds to reach full power from idle thrust - an eternity. This has caused aircraft to hit the ground if they have not been able to recover while in the down draught. Indeed, Boeing warns pilots that there are some down draughts which are not recoverable. They are so strong that they will out-perform the aircraft. That is in the most extreme conditions. To counter this, a lot of American airfields, especially in the Midwest where it is a common occurrence, have wind shear warnings which come from detectors on the ground near the runways. There are also wind shear warnings on the B767 and B757. It is indicated by a red light on the centre instrument panel, a red annunciation on both ADIs, and an aural alert - a two-tone siren followed by the automatic call of 'wind shear wind shear wind shear'.

Our pitch limit indicators (PLIs) on the ADIs indicate the pitch limit for stick shaker activation point for existing flight conditions. The PLIs show the maximum pitch to achieve maximum lift. Beyond that, we will stall, the wing will stall and will immediately lose lift. During all of this, there will be lots of turbulence, which makes flying the aircraft accurately very difficult. The aircraft will also be out of trim and the stick loads will be higher, again making accurate flying difficult. Because the PLIs indicate the pitch limit for stick shaker activation point for existing flight conditions, the PLI feathers on the ADI are constantly moving up and down. The whole process is incredibly dynamic but the pilot has to fly accurately and smoothly. You are flying in very rough conditions with the aural warning shouting in the background and the 'brrrrr' of the vibrating stick shaker on the control column. You will have used all your excess energy to give you lift and then all you can do is fly at that high nose attitude with full power.

On an approach into Bangor Maine in the States to pick up fuel on the way to Orlando, the weather was clear but very windy and as we turned right base on to final approach, we encountered wind shear. It was momentary and expected. We were not close to the ground and the warning was due to very strong winds over the terrain and not a thunderstorm, so I continued to land. I knew I could go around but the same could easily happen on the next approach. Boston, my alternate, was giving wind shear as well. We landed and it was okay, but as with all things, it was a judgment call.

The default situation will always be to throw it away because you never quite know how bad the wind shear is going to be. Judgment is such a critical aspect of flying - and judgment is all about being prepared for whatever conditions are thrown at you. If you encounter wind shear on the approach and have to go around, you will not be doing a standard go-around. With a standard go-around, once you have pushed the 'go around' switches, increased the power and pitched up, you will bring the flaps to a mid-position and then bring the wheels up. You don't do that in a wind shear situation. You leave everything as it is because maybe, just maybe, you are going to hit the ground hard whatever you do. If that happens, at least the wheels might take a bit of sting out of the landing. The only thing you would retract are the speed brakes. The point is that you don't actually know where the down draught is and you don't know where it is going to be in a minute's time. You have to accept that you are moving into a completely defensive position, very much as you would be in a pull-up, where it looked like you were going to hit the ground. This might happen if you are in cloud or at night, have lost situational awareness, and you don't know you are close to the ground. The terrain avoidance ground proximity warning system (EGPWS) can give cautions or warnings. One of the warnings begins with the aural alert 'TERRAIN, TERRAIN, PULL UP', red warning lights and messages, and you have to immediately accomplish the recovery manoeuvre.

Boeing clarified and emphasised the requirement to retract the speed brakes as part of the recovery manoeuvre after an American Airlines B757 crashed in Colombia, South America. Twelve seconds before they hit the hill, they received an EGPWS warning to pull up. Due to confusion with the flight management computer database, the crew selected the wrong waypoint into the system and under high workload and a loss of situational awareness, the aircraft turned in the wrong direction. By the time the error was detected, the aircraft was in the wrong valley and descending. Within a second of the warning, the crew carried out the escape manoeuvre but did not retract the speed brakes and the aircraft struck trees on the top of the mountain. The final report stated that if the speed brakes had been retracted during the escape manoeuvre, the aircraft could have been 150 feet above initial impact point. It might have cleared the trees.

Again, workload situational awareness and sound judgment are key elements to a good outcome. Judgment is based on knowledge and experience and on one's own attitude to risk. If ever I taxied out and heard someone mention wind shear on the RT, whether a pilot or ATC, I immediately clocked it and asked for full take-off power to be selected. Boeing says that if you have reported wind shear on take-off, you use full rated take-off thrust. It is as simple as that. I was nearly always the one on the flight deck who noticed it first. I have always been quite risk-averse, although that might not have been too obvious when I was flying little airplanes as a younger pilot. I have been able to judge where the line is, the line I would not cross in normal operations. Other pilots I have seen either back away from my line or, more seriously, go past it.

Of course, it is true that if the passengers don't get to their destination airport on time, they are going to be hacked off. But not half as hacked off as they would be if they all end up hitting the hill.

I always felt that some people didn't use sound judgment and that sometimes risk assessment was flawed. One person might come up with a good idea, then someone else built on it. The final plan would be better than any single person could have thought of. This is called synergy - but the opposite is risky shift, and avoiding risky shift is crucial to a good decision. Someone tentatively puts forward a flawed idea and another crew member thinks: 'I am not sure that was a good idea.' Nevertheless, they support the plan, thinking: 'It must be okay because he thought of it.' The spiral downward to a poor plan is on its way. There was an incident with a DC-10, which fortunately was not fatal, and afterwards the three chaps in the flight deck gave their account of what happened. It became apparent that one of them had suggested a course of action, a poor course of action. Since it was put forward by another crew member, they built on the idea, and by the time the idea had gone around the flight deck a few times, they were all convinced it was the way forward. Even after they had survived the crunch, they were saying: 'I thought it was a good idea...' They all bought into it - and they were all wrong. If you took them all individually before they had collectively bought into the bad idea and asked if they really thought it was a good one, the individual answers would come back: 'No, I don't!' The near-tragic outcome proved their collective decision-making was already part of the downward spiral.

There is another sometimes difficult aspect to turbulence which pilots have to constantly guard against, particularly in mountainous terrain. I have flown straight into something called a 'standing wave', which can occur when crossing over a mountain range such as the Pyrenees or the Alps. The winds aloft will not be running along the mountain range but over it. This effect can be experienced a hundred miles before you reach the area or even after you have crossed the mountain range and at great height. There are clues - lenticular cloud which is up at your level, a long cloud shaped a bit like the top of a mushroom or like ripples on a beach - and if you see one of those, chances are you are going to be running into a standing wave. Depending on the wind direction in relation to the mountain range, you can sometimes work out if you will encounter lift or sink first. Nothing to do with meteorology is a given, but the first thing you will see is an increase or a decrease of Mach number as you run into the wave. If the wind is a head wind, as you approach the mountain you would expect a loss of speed as the aircraft pitches up to remain in level flight. The auto throttle would increase the power but possibly not quickly enough and, in some cases, maximum continuous thrust will not be enough to keep you at your flight level. As you pass through the wave, the opposite happens and the thrust is reduced to prevent an over speed past maximum mach number (MMO) – speed reducing and power on, then speed increasing and power off. It can go on for 15 minutes with very large power changes.

As a result of standing wave, we have had a couple of over speeds and under speeds with stick shaker. Boeing has emphasised the need to use speed brake rather than just reducing power in an over speed situation. You do not want to close the thrust levers because within seconds you might need full power. Other than the clue in the cloud formation, it is invisible. In that situation, the air speed indicator is really all you have, plus possibly the 'seat of your pants' feeling that the air is not stable and the aircraft will be trying to hold the altitude and level flight. Some 'hot spots' where this is likely to occur are known. If we were crossing the Pyrenees, our flight plans would give us shear levels of turbulence which are the expected turbulence levels. We would also look at the wind speed and direction. If it was coming from the west along the Pyrenees, it was not so bad. Usually, if you see the wind is at 80 knots or 100 knots over mountains, you're going to be thinking about this possibility. And if it's night, you're not going to see the damn thing anyway.

With light aircraft, it is even more of a factor if you go mountain flying. I couldn't just take my Cirrus into some of the French or Swiss alpine airfields. Some of them need a clearance flight with an instructor, and with good reason because you need to have the local knowledge, just as we did when flying into Madeira. So I have only flown around the edges of the Alps in a light aircraft. On a beautiful day, you can fly down some of the more accessible passes but you need the right weather. Caution is always a good trait in a pilot.

From this day forward

The future of aviation - and improvements in operational standards, pilot competence and safety - is my passion. One area which continues to intrigue me is task management on the flight deck. Sitting in the back of a simulator is a privilege. You get to see the good pilot operate the aircraft in demanding and complex situations, and also deal with other crew members and analyse problems I have given them, to sort out the aircraft and keep it safe. They must communicate with various people and get the jet back on the ground. I have witnessed less capable pilots struggle to take control and manage the situation. More often than not, their aircraft handling is okay but their thought process is not clear or structured.

If they have been given a complex problem, such as a double hydraulic failure, the aircraft can become more difficult to fly and in some cases the autopilot will not work satisfactorily. One pilot needs to hand-fly the aircraft (PF, pilot flying) and get control of the aircraft, and the other (PM, pilot monitoring) reads the drill from the QRH (quick reference handbook). By the time the initial actions of the QRH are complete, the problem will have been contained and there will be a list of unserviceable equipment and possible consequences. In a double hydraulic failure, what they have lost will depend on which two of the three systems have failed. After switching off various pumps and identifying system pressure loss, it might indicate that they have lost some flight controls, normal flap extension, or normal gear extension. The next checklists are the descent, approach and landing checks, just before they make their final approach to land.

The bit I have seen them struggle with is from completing the initial

QRH drills to starting their approach and the final checklists. Imagine they have taken off from Manchester and, at 6,000 feet in the climb, the left and centre hydraulic systems fail with a total loss of fluid, possibly as a result of a tyre tread shredding and hitting hydraulic lines in the wheel well as the gear comes up. They tell ATC they want to stop the climb and take up a hold while they sort out the problem. The passengers and crew may or may not have heard anything. The QRH will have been completed and they know they are not going on to their destination. They are going to have to divert to the nearest suitable airport - but now the difficult part begins. They need a plan and there are hundreds of different options.

Over the years of watching many crews struggle, especially inexperienced first officers, I came up with a classroom briefing that would give them a checklist for any emergency. The most important thing is to fly the aircraft, in short ANC (aviate, navigate and communicate). There is no point in hitting the hill while making the next part of the plan. How urgent is the problem? If the aircraft is on fire or running out of fuel, there is little time to lose. They have to get the aircraft on the ground, nothing else matters. Even if they go off the end of the runway, it is better than falling out of the sky. Given a less critical situation, when time is not the major factor, they will need to speak to ATC, cabin crew, passengers, engineering, and the agent at the airport, and decide whether to declare a Mayday or a Pan Call. What additional checklists are needed if, for example, we are landing on a flat tyre? Have any circuit breakers popped? Do we need to cross-feed the fuel system? What is the minimum safe flight level (ANC)? Are we at the correct speed for the current flap setting (ANC)? Do ATC know what we are doing (ANC)? Then they need to gather information – runway lengths, what are the weather conditions like? Are the let-down aids serviceable? Can we land with all our flaps? Can we get the gear down? Can we raise the flaps or gear if we have to go around? Who is going to fly the aircraft while we sort all this out? Who is going to fly the approach and land? Do we get the cabin manager into the flight deck to brief or do we use the intercom? Do I need the runway of my choice sterilised so no other aircraft can use it? Small wonder that some pilots would be thinking: 'Where do I start?'

I would talk through all the possibilities, covering various failures by asking them to think of anything in any emergency situation that they MAY want to do before landing. Slowly, the board would fill up around the middle, which would say 'Aviate-Navigate- Communicate' or 'FLY THE AIRCRAFT'. Once they had a list of all the things they might want to do, given the urgency, they could get this checklist out and look at it. This would help their stress level because they wouldn't be floundering around grasping at straws. Former prime minister Harold Wilson smoked a pipe, and he said if an interviewer asked him an awkward question, he would fiddle with his pipe and light it, all the while thinking of his answer. In the same way, the checklist was to give the crew the opportunity to take a minute to look at the list and think: 'We need to do these things, but not those.' A plan would be coming together.

Another important advance in recent years has been the development of what has come to be called the 'Honest Broker' principle. I'm sure it will continue to be developed and improved. With the introduction of new technology, the company wanted to retro-fit a 'flight deck monitoring system' to the B767 and B757 in the mid-90s. They approached BALPA and I was on the company council at the time. There is still the 'Black Box' (flight data recorder) and the Pilot Voice Recorder, both of which are legally tied up and can only be interrogated after a major incident. Usually the pilots are no longer around by then. This monitoring system would continuously monitor hundreds of different parameters and the information used to improve flight safety. For example, if the data showed that pilots were finding a particular approach difficult to be stable by the required 500 feet, this information could be fed back to the authorities and the approach changed. Initially, it was seen as the spy in the cab. Pilots are happy with the Black Box and Voice Recorder but the thought of management pulling the data after each flight was not acceptable.

I was one of two people tasked by the company council to come up with something the pilots would accept. We visited BA, where the system had been fitted and accepted by their company council. After some heated discussion with the company, it was agreed that a new post of Honest Broker be established. Two or three pilots, who were acceptable to both the company and BALPA, would be rostered and they would go to Luton a couple of times a week to go through the data the engineers had downloaded from the aircraft. The crucial thing was that the data was confidential and only the Honest Broker and the captain would discuss any problem that came to light. The captain had the right not to engage with the Honest Broker but the vast majority were willing to discuss any issue.

I have had a few phone calls from the Honest Broker over the years. I was making an approach into Madeira in gusty wind conditions and just as I was touching down, I felt the aircraft suddenly become unstable. I fire-walled the engines and lifted off into a go-around. The second approach was fine and I landed. A week later, one of the Honest Brokers phoned me to chat about the approach. He thought I might be interested to know that the wind had changed direction by 180 degrees just as I was trying to touch down, which was why I was struggling to land. On another occasion, I was flying out of Stockholm, line-training a Scandinavian first officer on a flight down to Phuket. He was quite experienced on the B737 and had previously flown the B767. As we descended through 10,000 feet, I felt we were high and fast with a possible short cut to come. Sure enough, we were given a shorter route and were now obviously high and fast. The first officer took some action to get rid of the excess energy but not very aggressively. He said: 'This is the way we always make this approach.' As we descended through 1,000 feet, it was absolutely clear we were never going to get in, and there was no way we were going to be stable by 500 feet. I think he thought I was being a bit of an old fusspot and he was going to show me the way they did it in Scandinavia.

Sometimes you have to let them see for themselves that things are not going to work. We passed 500 feet and the landing flap wasn't down but he just carried on. As we passed 300 feet, I called 'go around' - and he did. He flew a visual circuit and on the second approach he was still a little fast. I told him what I wanted him to do to correct that. As we taxied in, ATC asked why we went around and I just said we were not stable. The first officer wasn't amused and said that if it were a Scandinavian crew, they would have landed. We had a little chat. When I got back to the UK, the Honest Broker phoned to ask why I went below 500 feet and continued the approach. I told him it was a training flight and the first officer couldn't accept that we were too fast. I said I had already made the decision to go around, back at 800 feet, but was demonstrating that the aircraft was never going to be stable. He just said: 'Thanks, Scott!'

One final story no one knows about happened when we were still negotiating with the company on how to introduce the system. We were looking at a flight into Palma by one of our B757s and what we saw was unbelievable. The aircraft landed on runway 24, which is a long runway. It had been so fast and unstable that only after touching down on the runway, closing the thrust levers, raising the speed brakes and starting wheel braking did the aircraft slow down enough for the remaining flap to automatically run to the full landing position.

Of course, it's not just on landing that pilots have problems. Some find themselves in trouble before they've even taken off, and aircraft have crashed trying to stop on the remaining length of runway after an aborted take-off. This was because the pilot chose to keep the problem on the ground, rather than take the problem into the air. In the simulator, the rejected take-off (RTO) manoeuvre is examined every six months and, of all the different drills, the RTO is the one which is failed the most.

The actual RTO is always performed by the captain, irrespective of who is doing the take-off. The decision speed (V1) is the highest speed by which the pilot should have made the decision to reject a take-off in the event of a relevant failure. There are two safety margins built into the V1 speed: recognition delay prior to V1, and a reaction delay after V1. The combined delay is one to two seconds. If you try to stop after V1, you will not stop on the runway. The total energy that must be dissipated during an RTO is proportional to the square of the aeroplane velocity. Therefore, at low speeds up to 80 knots the energy level is low and we can stop for any reason we feel is appropriate. But after 80 knots and up to V1, we stop only for a confirmed engine failure, any fire warning, wind shear warning, or if the captain thinks the aircraft just won't fly. So the decision to reject must be made by the captain before V1 to be able to initiate the reject drills by V1. This is made clear by the captain removing his hands from the thrust levers just before V1.

If you continue with the take-off before V1, the aircraft will not get airborne or, if airborne, it will not clear the obstacles in the climb-out path. Thus, of all the different drills and procedures pilots deal with on each flight, this is the most critical – and, in the sim, the procedure which is failed the most.

Style and luxury eight miles high

I have often been asked about the changing state of the aviation industry – sadly, I have yet to come up with a coherent answer. Aviation has evolved spectacularly since 1903 and the Wright brothers. No one is still alive who might have watched the Wright brothers fly, but many still remember the early jet era and the arrival of something like the Boeing 707. Only a short time after that, NASA was off to the moon. I am pretty sure that in those days people thought we would be on Mars by now and that we might even be colonising the moon.

In civil aviation, aircraft have reached a level of sophistication that few could have dreamt of. Tremendous progress has been made in the years since the arrival in the 1960s of the B747 Jumbo, which was really just a big B707 and technically not very different. The next generation of jets was far more impressive, including the B757 and B767. Then the Airbus came into service and we were introduced to the widespread use of advanced features such as fly by wire. Aircraft have become more economic to operate, using less fuel than before. I saw an old DC-10 in Birmingham on a kind of farewell tour. The DC-10 might have been a thirsty old beast but when fuel became cheap, companies could lease them for little outlay so they could return to operation and still make money. As soon as the fuel price went up again, they all found their way back to the desert. Now, most of the old gas-guzzlers have gone. You don't see the TriStars any more and you don't see VC10s. Even the RAF has got rid of them. And, of course, the Comets are long gone. The B707s have gone and the B727s have almost disappeared in America, where they were once such a reliable workhorse.

What we are left with is an industry full of reasonably modern jets which can be found almost everywhere you go in the world. The latest generation Airbus A380, Airbus A350, Boeing 787 and the new B747 800 have improved levels of sophistication which might include electric brakes instead of hydraulic brakes, which means they don't have to carry 80 gallons of hydraulic fluid, as you do in a B767. What surprises me is that most airliners' maximum altitude is still just above 40,000 feet. The B787 will get up there quicker and easier than, say, a B767. Over the ocean, we usually started westbound at 33,000 to 35,000 feet, whereas the B787 will go straight up to 38,000 feet, which gives it a bit more operational freedom. Only military and private jets fly above 43,000 feet. Obviously, Boeing and Airbus don't think that is worthwhile. Modern jets with modern wings will always be able to fly a little bit faster, a little bit higher and, most importantly, more efficiently. We were flying at about Mach .80 (80% of the speed of sound). Later aircraft will fly at about Mach .84 but even the Jumbo flew at about .85. In terms of flying time to the Caribbean from the UK, the latest aircraft will get there half an hour earlier – which, all things considered, is not really a big gain, although they will get there by burning a few tonnes less fuel.

There has not been a great leap forward, fundamentally because of the cost. Concorde's supersonic engines were military ones and extremely noisy and fuel-inefficient. Bringing out a jet capable of going supersonic is a big problem. That is why they are now looking at hypersonic flight, Mach 5 and above, so it might be possible to fly from London to Australia in four hours. To do that, they will be flying outside of the atmosphere. People like Richard Branson are pushing the boundaries, but building a quiet supersonic airliner is very difficult. It's not insurmountable but the costs are colossal. Airlines want something that's going to make them money. Concorde could not make money because it carried only 92 to 100 passengers and didn't have very good range. No country wanted Concorde flying overhead, creating a supersonic boom, so it was a magnificent white elephant.

Jets have got safer and more reliable but they are still just jets. The cost to the passengers has come down and down. Everyone is off everywhere, either for business or for pleasure. The chances are that, sooner or later, fuel is going to go up in price, and that will make a huge difference to the price of a holiday. Concorde was conceived before the huge fuel price hike in the 70s. If something like that happened now, you would find that of the 20,000-odd planes that are airborne at any one time today, a lot will be on the ground - and 20,000 aircraft are burning approximately 80,000 tonnes of fuel per hour. The next fuel crisis will have airlines falling out of the sky, as they did when airlines such as Swiss Air went out of business. Who would have ever thought an established and reputable airline such as Swiss Air would ever go bust?

The nicest plane I have ever flown is the B767-200, the one with the big engines. It climbs, as they say, 'like a homesick angel'. It has inboard and outboard ailerons, unlike the B757, so it has a great roll rate. I could usually tell if someone flew the B767 occasionally and was more used to the B757. The same thing happened if I hadn't flown a B767 for a while. You come down the ILS rolling because the ailerons are very powerful and therefore sensitive. I always had to say to myself: 'Take your hands off the controls!' Then the self-induced pilot roll stopped. The B747-100 and 200 flew just like a B767 but like all four-engined aircraft, was gutless compared to a two-engined aircraft. But the B767 is a beautiful aircraft to fly.

I don't have great sentiment for the Jodel I owned but I do for the Piper Cherokee, the Ford Escort of the air. It's been around a long time. There are not many small seaplanes like the one I landed on Loch Lomond all those years ago, the Lake Buccaneer, but I guess something from even earlier that still makes me go 'Oh yes!' would be the single Otter, or the Beaver which was its little brother, both built by De Havilland. I would love a go in both of these some day. I flew the Twin Otter when I was in Loganair but the single Otter was a classic 1950s aircraft, a great big brute of a thing. Of course, like everyone, I look up when I hear the wonderful sound of the Rolls-Royce Merlin engine and watch a Lancaster, a Spitfire or a Hurricane and think: 'Bloody marvellous!' Other military aircraft have never really done much for me. My passion has always been for the civilian aircraft. Even today, there are not many aircraft that fly over my head where I don't glance up just to identify it.

Mentoring – as it's meant to be

Mentoring doesn't exist as a natural concept for the majority of captains. When I look back at my training and my experience, I learnt a lot in the simulator, the set pieces, the dance steps for command. But the style and the panache required for command you can't expect to learn just in the simulator. Over the years, I have learned from the old captains, both how not to do it and how to do it. Whenever I spoke to captains about any particular aspect, they would tell me: 'That's how you do it.' I was able to pick the good things and leave behind the bad stuff. During my early days, the authority gradient in the flight deck was much steeper than it is now. I wouldn't have been able to have a constructive conversation with all of them - but I did watch them very closely.

Today the authority gradient is much flatter and things are usually much easier-going on the flight deck. In my view, this is feeding into a culture where the imparting of knowledge in that environment is just not strong enough. Mentoring by the captains helps develop both seats. If I sat on most flight decks as an invisible person, the chat between captain and first officer would be 90 per cent about things other than the serious business of flying. Of course, there would be a bit more homing in of attention when they started briefing about their descent, approach and landing.

We did have mentoring by certain training captains to first officers who were coming up for command, which was good if you happened to have the right training captain and the opportunity of going away for a few days with the same trainer, with time to chat informally over dinner about all sorts of flying topics. The big question is: how can you ensure that captains mentor co-pilots as a norm, little and often, and in a fairly low-key way? It would be of great benefit to young pilots and to the industry as a whole if this could be resolved correctly. It needs to be introduced slowly, so that in five years' time, the captains are confident enough to pass on their very valuable experience on the flight deck. Perhaps a good start would be a select group of captains all of one mind on this matter. Flying tips in real-life situations are invaluable and cannot be replicated in the classroom or the sim.

There could be a downside to this. I promoted a captain quite a few years ago who ended up causing chaos on the line because he was all over the first officers like a rash, asking them question after question, and really hacking them off. When I had a word with him about his approach, he replied: 'I'm just trying to pass on my knowledge to bring them on.' The sad part was that he had no clue how and when to do it. He eventually left the airline but not before he had caused a lot of problems.

Unfortunately, as with all things, there are some people who will get it and there are a few who do not. Some will say: 'It's not my job. I'm not a training captain.' In a situation like that, you just have to accept it. Whatever system they come up with, it should be robust enough that if people don't want to make this extra effort on a co-pilot's behalf, there is no compulsion for them to do so. But any improvement to the current situation will be a significant one. Imagine what people could learn from the captain who ditched in the Hudson. What was his thinking process which led to such a good outcome? Why was he successful when many wouldn't have been? In the long term, it would be self-sustaining, because a mentored first officer would gain a benefit from it and then mentor his co-pilots when he became a captain. Both the captains and the co-pilots have to engage in the process. The co-pilots have to be hungry to learn and develop, and it should be the norm that captains are happy to be part of the process. The first officer who actively questions his colleagues and searches for answers will probably achieve his command in a shorter time because he will be better prepared for it. He would look like a captain before he even started the command course. In an ideal world, mentoring might one day be the norm.

There are good captains out there who don't want to be training captains who have a lot to offer their colleagues. However, training captains are instructed and developed in the skills of training. If the other pilot is hand-flying the jet, there is little point in giving them some top tips since they will be too busy. They can't listen properly. But there are many high-calibre men and women who would be able to give very sensible and practical tips on the job. I was told by many captains, when I was in the right-hand seat, to take time in the cruise to consider, and then have an active dialogue with them about, all the 'what if' scenarios throughout any particular flight. It would have to be done gently and not all of the time because you wouldn't want somebody knackered by constant questioning. There would be great benefit in the captain asking his first officer, quietly: 'Have you thought about this?' If they said yes, they could be told: 'That's great. Now let's take it a bit further.' There are only so many things a jet can do and it is important to talk through the various 'what if' scenarios, such as, for example, depressurisation, engine failures, hydraulic failures - or multiple hydraulic failures. But all this can be done in a non-threatening environment. I really believe it is worth doing and I carried this out very successfully myself over many years.

Getting an initiative like this off the ground needs someone with great CRM skills who would be able to develop the concept so it can be introduced in a wholly positive manner. Input from the first officers would be crucial to achieve a good outcome. The short-term commerciality of an initiative like this should not become an impediment. Compliance in aviation is the norm, but compliance is actually a race to the bottom. In the Britannia days, we wanted our standards to be better than compliance, which was meant to be the minimum standard below which it was not acceptable to fall. The Authority's role is just to ensure compliance, to say: 'That is the minimum standard and if you don't meet it, you don't fly the jet.' This would apply to the people in engineering as much as to the pilots. One should aspire to more than that. Compliance has to be realistically at the lower end of what is acceptable.

Our first chief training captain's brief was to take this ragbag of a post-war charter airline and turn it into a world-class operation. In today's world, 'charter airline' is not the dirty word it was in the 1950s. He made the airline world-class and he began that process by demanding high standards. Pilots used to be tested and scored on everything from communication to hand-flying the jet and technical knowledge, on a scale of one to five. One was a fail, two a pass for the first officer, but a fail for the captain. Three was command standard. Four was above standard and five was high standard. There was a difference between the minimum standard for a first officer and a captain because we believed the minimum standard for a captain must always be higher. On a nasty, windy night, the captain does the landing. The Authority doesn't see it like that. In the Authority's world, two would be a pass for a captain as well.

Training captains frequently said: 'The minimum standard is that I am happy to put my wife and children on this captain's jet at any time.' People would say: 'Well, that's what compliance is.' But it isn't, because compliance has to be written for the weakest acceptable airlines in the country. All airlines should strive to do better, so a concept like mentoring becomes vitally important, and mentoring is nothing to do with compliance. Mentoring is what good looks like - compliance is not.

A virtuous circle of caring

Compliance is intended to support standards within all airlines, but that is not always the case. Some UK airlines have reduced the yearly pilot simulator training and testing by four hours, although that is still regarded by EASA and the CAA as compliant. Most pilots would rather have those extra hours, even though they don't particularly enjoy being tested in the sim. They know it is the better way. Pilots often struggle with the single-engine work because we never actually fly the jet with a single engine as they are so well constructed and jet engines are incredibly reliable. Nevertheless, every six months we used to go into the sim and practise because it was important from a safety point of view. In some airlines that sim practice takes place only once a year, simply to save money. If I put the names of ten airlines on a desk and asked pilots which they thought were the better, safer airlines, the ones not just hanging on to compliance, people in the industry would know the answer straightaway. The public do not.

Even within Europe I would reluctantly get on some European carriers with my wife and kids. There was one particular carrier - bust now - which was a national carrier but which I would have avoided like the plague. I have trained pilots from airlines just outside Europe and I wouldn't get on their jet if they paid me. In fact, one of those airlines was once banned from UK airspace entirely. Sadly, I have to conclude that the words 'compliance', 'cost' and 'lip service' all have to come in the same sentence. Of course, I am only too aware that you need airlines to make money to be profitable but I do not believe that a drive to the bottom is the way forward. I do not believe airlines should be advocating single-engine taxi-out just to save a bucket of fuel per flight, especially when some aircraft manufacturers are changing their taxi procedures especially to reduce pilot workload while taxiing. Aircraft burn fuel, that's just the way it is. Flying economically makes good sense, I understand that, but introducing procedures to complicate an already busy and complex time of the operation is not sensible.

The one aspect of flying we don't do as an industry is to sell safety. There is always the big statement that every airline will talk about – 'safety is our number one priority'. Everyone says that. An aircraft from a large UK carrier had to divert recently - and the first thing the airline said was: 'Safety comes first.' It has become a mantra. The question is: 'Who really means it?' Of course it has to be safe, but that is the very minimum requirement. Compliance is cited as ticking that box. If it's compliant, it's safe. It is something of an unwritten rule that all aircraft are safe, but I have never heard any airline claim that they invest in more complex training than anyone else. The reason they don't say these things is because they don't. Airlines don't even take the basic precaution of suggesting to passengers on board their aircraft that they might avoid wearing flip-flops. In an aircraft evacuation, flip-flops will not prevent feet being cut to pieces while moving through the debris. We are in danger of turning the whole thing into a computer game.

In the 1960s and 1970s, the Japanese famously had a pyramid system of company management: the managing director was at the top and below him there would be all the layers of management, until you got down to the workers at the very bottom. That's the way top-down management has been set up. When they realised why that did not work, they came up with a completely different structure. They simply took the same pyramid and inverted it so the management would be at the bottom trying to support their people. I don't think either is right, although I am no businessman. It strikes me that you need a two-way structure. You need a process where the people at the top and people at the bottom have the support of each other and communications are not bogged down and fogged up by middle management.

There are some more enlightened companies – US ice cream firm Ben and Jerry's was one of them – that have adopted rules which say the managing director can only earn so many times that of the lowest salary. It indicates respect at all levels. The chap who owns the Timpson shoe repair chain is well known for saying that he quickly discovered that it was more efficient and profitable, and better for staff, if he just let the individual shops run themselves. Sir John Timpson says they are the people who really know what is needed, not head office. They trust their people to run the shops. According to him, the culture of large companies, in his experience, was entirely different. He said that if someone does something wrong – like dipping their hand in the till – the big companies try to plug the gap in the system by saying that no one could use the till unless there were three people standing over them, watching their every move, instead of looking at the individual who stole and telling them they were no longer wanted in the company.

When I visited the airline JetBlue in the US at their training centre in Orlando, what impressed me most was what one of the very senior management team had to say to all of the staff. He said they were the boys and girls on the front line. They were the ones, when everything was snowed in and the aircraft were grounded again, who have to deal with the huge queues checking in and passengers becoming really cross. This chap promised that he and his senior members on the board would be at the airport as soon as they possibly could and would immediately be at the side of the front-line troops under pressure. The people behind the desk would be able to say to any rebellious passenger: 'If you join that queue, you'll be able to speak to our managing director.' That is real support. That is the sort of company I'd want to work for. He inspired all who were there. He went on to say: 'It's the little things that count. I am not going to tell the flight attendants how many packets of nuts to give to each passenger. If you give out one or two and they want more, give them more - give them the whole box if that makes them happy. It costs us almost nothing.'

Ben and Jerry's used to give 7.5 per cent of its pre-tax profit to a foundation to help its people. Contrast that with the fate of our people who had to fight over an extra half of one per cent. The owners of inspired modern companies say the whole structure should actually be the other way around. They say you start with a decent product but then it is all about the staff. The aircrew need to be looked after first. They need to be committed, motivated and cared for, because if they do a good job, the customer experience is going to be good from the outset and passengers are going to feel well looked after. If the passengers are pleased, they come back and fly with you, and the shareholders will be happy. They will remember if the crew were surly or unhelpful. Sometimes that is the norm. That's an experience many people have had when they board an aircraft, when they see the crew hiding in the corner and looking bored while people struggle with bags. The reason for this is the crew do not feel they are cared for. It is a simple solution: if the company really care for the staff, they will be delighted to care for the passengers.

Some airlines are doing a good job but the pressure to reduce costs is on all airlines and many are being dragged down by what I call 'the bottom feeders'. These are airlines which are happy to take advantage of young men and women as pilot recruits after they have spent in excess of £100,000 to get a professional licence and a jet type rating. Having been through all of that, the pilots are so desperate for a job and so skint that airlines can do to them pretty much whatever they wish. Some of these airlines won't even let BALPA represent them so there is no union protection for these young and inexperienced pilots.

The fact that people have paid very large sums of money – usually borrowed money – for their licences is not new. It is quite normal. I had to do exactly the same, although the cost was nowhere near 100 grand. And back then, quite a few UK airlines sponsored new pilots through their initial professional training and then their aircraft type rating. Those who were not sponsored, like myself, were given a type rating paid for by the company. Now many new pilots are starting their careers owing tens of thousands of pounds, only to find poor terms and conditions when they do eventually join an airline. The spiralling down of pilots' terms and conditions and the mantra that we are compliant has a direct effect on flight safety. Added to that, some airlines have reduced the minimum hours required for command. When I went on my command course, the minimum requirement was 5,000 flying hours. I recently saw an advert for a captain with a UK airline which required 2,000 hours for a B737 command. Unbelievably, another airline was trying to go even below that. These people will be allowed to fly and command a complex jet airliner.

Most companies are pushing to get more for less. One long-haul carrier wanted not to include a pilot's time in the bunk as flying hours. When the captain was sleeping in the bunk during a very long flight, they didn't want that time counting to the 900-hour yearly total. Quite rightly, the Authority said no.

Then along came the EASA decree which said: 'You can now work 1,000 hours in twelve consecutive months.' Like the vast majority of pilots in this country, I could see no justification for that. BALPA had a huge campaign to stop the new EASA flight time regulations being adopted. They even lobbied the government, but to no avail. The suggestion that UK airlines would not roster to the maximum was ludicrous. Of course, no one asks the public if they want to be in the back of a jet with someone who has been awake for 20 hours and is now about to make his approach in grotty weather to a demanding landing. Passengers expect the captain flying their jet to be well-trained and experienced. A captain with 2,000 hours might have held a professional licence for only two years. They might have joined the airline with a couple of hundred hours from initial training and then in two years – possibly one winter – they could be sent to Lapland on a Santa flight as captain. I am certainly not getting into the back of that jet, because if the captain has only got a couple of thousand hours, what has the co-pilot got?

Sometimes a lack of safety features in aviation is down to the arrogance and short-sightedness of airlines and aircraft manufacturers. In 1992, a French airline's Airbus A320 crashed with considerable loss of life. The pilots had no warning of the imminent impact because the airline had not equipped its planes with a ground proximity warning system (GPWS) and the manufacturer had sold them the aircraft with no GPWS protection. The FAA in the US made GPWS mandatory in the mid-1970s following a spate of crashes, including a TWA flight at Washington and the L-1011 TriStar at Miami. The International Civil Aviation Organisation then recommended that GPWS be mandatory but the French did not adopt this until the early 1990s. Even the old B737-200s I flew in the early 1980s had a basic GPWS fitted.

Pilot Al Haynes, on that DC-10 which crashed at Sioux City in 1989, saved more than half the people on board. Everybody on that aircraft should have died and it was only the immense skill of the pilots and flight engineer which reduced the carnage. McDonnell Douglas, the jet manufacturer, had fitted General Electric CF6-6 engines and the number two engine – the tail-mounted one – had been manufactured with a flaw on the fan disc. When the engine exploded during the cruise, all three hydraulic system lines were severed. The only thing that saved half the people on board was the ability of Al Haynes, who had more than 29,000 hours - 7,000 of them on the DC10 - and the other flight crew to think out of the box.

As they were joined by a training captain who happened to be on the plane, the flight engineer was mostly talking to engineering, who could not believe the situation they were in. Al was trying to fly the jet through the controls, which were doing nothing because they had lost all the hydraulics and there was no manual reversion. They were trying to control pitch and roll by increasing and decreasing power on the remaining underwing engines. Effectively, they were trying to fly a brick which had two underslung engines on the wings and which was in a permanent right turn. Months later, when investigators went through the exact sequence of failures in the simulator, not a single simulator crew had a successful outcome. Through his ability to open the flight deck to a stranger, who happened to be another pilot, the three pilots and flight engineer worked together and came up with a kind of solution to the problem. Computers could not do that, and will never be able to.

I have sat in the back of simulators for thousands of hours watching poor decisions being made by a few mediocre pilots. One crew came into the debrief after their sim session and I asked: 'How did that go?' The first officer seemed pretty pleased with himself until I pointed out the one glaring mistake he hadn't picked up on – the jet he was flying and practising for his command on their simulated approach was almost out of fuel. Exceptional pilots, like the chap in the Hudson River, can save the day by being fully situationally aware and immediately understanding what had happened and why the best outcome would be achieved by throwing away all the airfield options and going for the river instead.

The way to achieve that high-calibre pilot is not easy – and it cannot be achieved by cutting corners. It might take additional investment to give all pilots the opportunity to be trained to that standard. They need to have done it long enough to gain the correct level of experience. Some people say: 'I have 10 years' experience.' But it's the same first year on the line experience for the next 10 years. They haven't gone on and built on their initial year and developed the CRM skills learnt on their apprenticeship for command. So we have to have experience, excellent training, good technical knowledge and a training environment where pilots can experience all the possible real-life situations they might need to deal with. That requires many simulator sessions, which you cannot achieve if you only have 2,000 hours. There needs to be training for the failures, which are incredibly rare. Loss of all engines, no flying controls and high-altitude jet upset can still happen, even though they are very rare events.

Civilian pilots never experience a real decompression in a decompression chamber, whereas Navy and RAF pilots do. In that environment, there is more chance of an explosive decompression - and they don't have a union. We practised decompression drills in the simulator. We started at cruise altitude, around 37,000 feet, and suddenly there was a 'bang' and a 'whoosh' and the aural warning and messages indicated a decompression. We immediately donned our oxygen masks and performed the drill from memory. It is incredibly useful training but, like many things within the simulator environment, it can only simulate reality to a point. A robust safety culture is supported by a good training department and a good management. It cannot be taken for granted and has to be reviewed and clearly understood by the workforce.

Getting back to basics

Let me paint a scenario which exposes the potential problems when departing with minimum required fuel and a day which is not going as planned. You're coming into the UK and, as a good company pilot, you have taken minimum flight planned fuel. Because the weather seemed to be nice, you were happy to just take minimum fuel. This would mean that if you couldn't get the flight level you wanted and ATC wanted you to start descent early, you would have burned most of your contingency planned fuel of about half a tonne. You would still be above company minimum reserves and therefore have enough fuel to fly to your destination, hold for half an hour, make an approach and go around, then fly to your diversion airport and land. So you are descending through 5,000 feet at your destination and you call: 'Flaps one.' Your colleague selects flap one and the next thing there is a flap warning message. You don't know what the problem is but you break off the approach and hold somewhere while you work the problem. There are many flap failure scenarios covered in the emergency QRH checklist and as you are working your way through the checklist, it becomes clear that none of the trailing edge or leading edge flaps are working. As a result, you will need to do an 'all flaps and slats up landing'. You have lots still to do, limited fuel and therefore limited time. You need to come up with a plan, brief the other pilot, brief ATC, brief the crew and the passengers.

You will need to check the landing distance required for an all flaps and slats up landing and complete all the required checks before landing. If your approach was into, say, Newcastle, where the runway is not overly long at 7,641 feet, your diversion fuel is usually for Teesside, where the runway is 7,516 feet, which of course is no better for you. Then you remember that Manchester's runway is 10,007 feet, but more than twice as far away as Teesside. You also possibly have some flap stuck out so you are speed-limited to 240kts. The day is not going so well now! You declare a Mayday and tell ATC you are diverting to Manchester and require radar vectoring direct for runway 23R, where you intend to land well below minimum fuel reserves.

I am not suggesting uplifting extra fuel for no reason. I guess I carried minimum fuel on about 70% of my flights. If there are no problems envisaged, then minimum fuel is appropriate. In my little scenario, the weather was okay but the pilots hadn't considered the fact that both their destination and alternate runways were not long. I always said that if my destination runway was short, my alternate runway was long, or vice versa. If one of the Newcastle-bound pilots had thought about the two runways, they might have elected Manchester as the number one alternate and then taken minimum fuel for the new plan.

When I was flying to Greece, there were only three airports with ILS (instrument landing system) in the whole country. If I was going anywhere, unless it was absolutely gorgeous weather, I made sure that either my destination or my alternate had an ILS, a precision approach facility. When I mentioned this to co-pilots, they tended to reply without ever having considered this fact beforehand: 'Good idea, Scott...' By not considering the possibility of aircraft failures, or deteriorating weather, or whatever, we give away our options, making subsequent decisions much more critical. We are forced into a corner. This is a lack of 'what if' preparation.

What if I have a double engine failure just after take-off? Where am I going to put the jet? That is what good commanders think about. They will often take minimum fuel but only after careful consideration and with the confidence that if the day doesn't go to plan and they have used more fuel than expected, they will be able to stop and pick up some more. That takes real confidence in your decision-making process.

We can do much better in preparing for the day it all goes wrong, which poses the question: 'How much better should it be?' I don't know the full answer to that, but in many airlines the standard is minimalist, which doesn't identify as a 'good safety culture'. Cost-cutting is loved by the shareholders and the City. Of course, indirectly the passengers like it too because they want the very cheapest flights and holidays, but when that is served up, they believe that the safety box is still well and truly ticked. The harsh truth is: it isn't. The box is ticked, compliance is achieved, but quality is missing. The public don't know that because it is complex and the airlines have lost their way. I have sat for thousands of hours in the flight deck with this constant thought: 'We can do better than this – we must do better than this.'

I have always regarded myself as an optimist and I still believe that over time there is a way to gently move the present culture forward to a point where the commander's knowledge, ability and judgment are at a higher standard to give the person sitting beside them the benefit of their vast experience. This should be the norm. I don't blame the captains. I blame the lack of a good robust safety culture. In Scott's airline, that would be quickly sorted out. If people in charge won't listen and respond to professional experienced common sense, they shouldn't remain in charge.

The minimalist approach to professional flying reached a new low when the authorities came out with a professional pilot's licence that does not require the commercial pilot to have flown in a small aircraft enough to achieve a PPL. My own flying history, like scores of others, began in light aircraft where we learned basic flying skills. Aircraft will always have an inherent ability to scare the 'bejeezus' out of the person at the controls, however good they think they are. I quickly learned that even little aircraft could kill me, which imbues a vast amount of respect for the physics of flying, even when we become cocooned in the cockpit of a modern jet which is presumed to be invincible. In my early years, flying light aircraft and as a flying instructor, I was a pilot with no discipline and with little sound judgment. I got the odd severe talking-to from Jim Beaton but, for whatever the reason, my immaturity was not knocked out of me. I have described a few near-misses with death while fooling around in a light aircraft and luck, more than anything else, saved my bacon. Only after joining my first airline was I introduced to a culture of discipline and safety. The airlines I joined gave flight safety great importance.

There is no doubt that advances in technology have made aviation much safer. Each decade, there are huge steps forward, making the aircraft safer, easier to operate and more reliable. Yet in the last few years a number of jet aircraft have been lost, not because of technical malfunction but because of pilot error. Pilot training and checking is not keeping the skies safe enough and relying on compliance is not producing the calibre of pilot we need in our modern airliners. I am not convinced new pilots coming into the industry are being nurtured enough. The advances in technology are masking a reduction in pilot skills. Most airlines discourage hand-flying the aircraft. Let's encourage every pilot to get back to practising the basic skills of hand-flying, at least now and again and when appropriate. After all, these are the very minimum skills the paying passengers presume the pilots possess. They are never told that the captain and his co-pilot are discouraged from hand-flying the jet at high altitude. To be fair, letting the autopilot fly the aircraft is more accurate and reduces the pilot flying's workload massively. In the simulator I have witnessed the pilot flying disconnect the autopilot at the onset of a problem and then seen their workload and stress level increase dramatically. The idea of pilots hand-flying as the norm is not what I am suggesting. On a busy departure, I would expect the autopilot to be engaged so both pilots can monitor the operation. But out of a quiet airport, in quiet airspace, I would say: 'Fill your boots!'

Passengers think we sit there flying the jet like they do in the movies. If we did, it would be exhausting and they would be airsick. We rely on computers and machines to fly the aircraft accurately, allowing us to monitor and manage the flight. But if a pilot is confident in hand-flying the jet when technology does let him down, he will be more comfortable in the situation he finds himself in. Sitting cosy at 38,000 feet at 550mph, letting the automatics take you to your destination could make the pilot remote from the jet. Every pilot needs to get back occasionally to the simple way of flying – to think about their instrument scan, what attitude is required to keep level flight, and to understand the power setting required. All the information is in the quick reference handbook, but it should be one of the many scenarios thought about during a quiet time in the cruise. What we always have to consider is that when everything is going wrong and you have lost all the fancy technology to run the show, what then? You are back to basics. Now what are you going to do?

The mental health of pilots is not spoken about or even measured within the world of commercial aviation. I have known three captains in my airline who have killed themselves and pilots who have resigned through mental health issues. It is an area which has been almost totally ignored in aviation, until very recently. The latest interviewing techniques may have changed but until fairly recently, as long as the candidate had a Class One medical certificate, the health box was ticked. We have no real idea of the pilot's mental state. The only time in my whole career that I came across a psychologist was when I was becoming a senior examiner. The question of pilot mental health has arisen around theories about the fate of the Malaysian MH370 jet which disappeared in the southern hemisphere. There have been documented examples of pilots killing themselves by crashing their aircraft, but very few. I don't know what we could do about it if we did find out what exactly happened. It might transpire that whatever happened, the pilots would have passed all the tests thrown at them. What I do know is that the area of mental health in pilots is never discussed. When I was the boss of pilot training for our airline, I said that if the day was not going well in the sim and a candidate was really stressed, ask them if they have a personal problem. The Authority does say clearly that when you pitch up for a simulator check, you should be in the same physical and mental state you would be in as you climb on to an aircraft. Over the years, I've sent a few people home after this discussion without failing them. That is no longer the way it is handled. Now they go home failed. It is unlikely that a colleague will recognise a problem in the demeanour of a pilot which might cause concern, as often the pilots have never met before or at least never flown together.

In 2015, a jet airliner crashed in Europe with the loss of all lives on board and this time the co-pilot was intent on killing himself and everyone else. This was not the first instance of an aircraft full of men, women and children being used by a pilot to end their own life and probably won't be the last. The media was full of discussions on what to do to 'stop this ever happening again'. Sadly, that might not be possible. The general consensus seemed to be to have a cabin crew member in the flight deck when one of the pilots is out having a comfort break. But a non-trained cabin crew member wouldn't be able to tell if the remaining pilot was descending the aircraft inappropriately until the EGPWS (enhanced ground proximity warning system) activated. The cabin manager could train in the simulator to know what to look for. Better still, include a third qualified pilot. The problem is, of course, cost.

When I flew with one of the captains who took his own life not long after, I had a lovely day out with him. There was no sign of anything untoward. I have had more than 70 medicals over my career and never once have I thought my mental health was being examined. If they want to try to make these terrible events less likely, the authorities, together with medical experts, should develop psychological screening for all new pilots and in yearly medicals. All pilots being promoted to captain should be screened. This might possibly make a difference but it will cost. It is so, so easy to pay lip-service to this latest tragic dimension to civil aviation. The bottom line is any pilot on any flight can destroy their aircraft and all the human beings transported in the cabin. The public are now aware of a list of aircraft brought down by the person who should have been there to save them.

Pilots need to be looked after, respected for the role they do. A fully functioning safety culture should support pilots when they genuinely feel there is, or could be, a problem. I believe my airline was better than most in this regard. If I phoned up Crewing to say I didn't feel well with a head cold, I was off the flight. The weakness is that it is up to the pilot to say something to either the company or to BALPA if they are struggling with their mental health. Pilots have to be able to admit to themselves that there is a problem. However, they have to believe that if they are struggling at home with health or with bad news, they will be supported. I know a couple of pilots who had a drink problem and the company sent them on a programme to help sort themselves out. That's important but I don't believe that is across the board with all airlines. This confidence - that the pilot will be looked after sympathetically - is a fundamental part of a good safety culture.

As for the EASA flight time limitation rules, I've not seen anything that is going to make rosters less onerous. All pilots know they can work a hard roster if it's a well-written one. This means they will not be going from lates, departing the UK in the evening and arriving back as dawn is breaking, and then have a blank day before moving on to earlies where they will be getting up at 4am. There have to be enough pilots employed to make the flying programme work. If they have a stable and well-written roster, the transition from lates to earlies can be achieved gradually and sensibly. If we had ATC delays or a technical delay, we could work a 17-hour day. You would never expect a surgeon to work 17 hours and then to make his final cut at the end of that time. But when it comes to the most important thing - the final landing - pilots are expected to pull that off perfectly at the end of a long day.

Manifesto for a modern airline

The quality and vision of management has not improved over the years - most probably, it has got worse. If someone said to me: 'Scott, we'd like you to start an airline and want you to develop a healthy culture within the aircrew and open that up to all the workforce', I would give that very serious consideration. I would love to be part of a caring, enabling, trusting culture and a management which inspires passion within the workplace; a place where the staff truly feel cared for. The results would be fantastic: the passengers would see and feel the difference and we would be head and shoulders above the competition. If it was Scott Birrell's airline, the financial bottom line would only be part of the story. We would need to be profitable but profit would not be the only way of identifying a successful company. A workforce would know that, in times of need, the company would be there to support them.

Years ago, when Ruth was ill, the company could not have done more for us. The support was fantastic and we have never forgotten how they looked after us. Sadly, in later years, one of our captains contacted me for advice about the company's private medical health scheme. His wife was not well and he felt he was getting the run-around from the company. I spoke to our management who assured me all was well - but he certainly didn't see it that way. This lack of care causes a lot of concern within the pilot workgroup.

In Scott's airline, against any financial backdrop, there would be a company statement which would recognise our responsibility to our workers, to our passengers and to the City – in that order. An airline is not run just for the City. Surely, the people who really matter are the workforce? There was a boss of a large UK company who was happy to share her home telephone number with staff and customers, demonstrating real support for both. This lady was very open about the fact that she believes her staff are everything to her and the success of her business. So is there an airline that has got it right? JetBlue certainly looked like it was establishing a caring culture when I visited in 2005. Virgin was on the right road during the time I was there, certainly as far as the cabin crews were concerned. They thought Richard Branson was magnificent. They weren't at all overpaid but he made sure he looked after them in other ways. I am not saying everything was perfect at Virgin, by any means, but there was a real feeling, whether it was deserved or not, that we, the crew, were the best. A lot of attention was given to appearances. The aircraft when I was there tended to be mostly old dogs, but despite their age, Branson insisted that they looked brilliant from the outside and fabulous inside. The onboard service was designed to be buzzy and different.

Of course, the company was led by a high-profile celebrity and it was his new train set. Twenty years after I left, in 2012, I flew back from Orlando in the back of a Virgin flight and the experience was not as good as Thomson. The seat pitch, the food and the service were appalling. Whether it was a different world in the flight deck I don't know, but in the cabin that airline had definitely lost its mojo. The crew didn't smile. They were like any other in any other airline anywhere in the world. I know more than most that it's hard to smile through a long overnight flight, but crews who feel appreciated might manage the odd smile. You have to motivate, you have to empower your people. There is a hotel in London where the staff can offer a gift or a refund to guests if something has gone wrong - without asking their management. It applies to all staff, so it could be housekeeping or the room cleaner. The point is they are giving empowerment and then following up with further training. Some people find empowerment very difficult and businesses that adopt this policy do find a minority of staff can't cope and need to move on. This is most often identified at the management level, where some people can't deal with managing empowered staff.

In our business, one of the biggest things the passengers liked when there was a delay was a free bar service. The alcoholic drinks we sold on the aeroplane made us a lot of money. It was a very important part of our revenue and paid the salaries of most of our cabin crew. There have always been inevitable delays and diversions. My view was that if we had to divert somewhere because the weather was bad, I would tell the crew not to give free drinks to passengers because it was not our fault. However, if we had a technical problem and were delayed, I would ask the cabin manager to offer a free drink to all passengers because it was our fault. These days, cabin managers are under more and more pressure not to do so. If they do, they'll find themselves having to justify their decision. A few years ago, I fancied a diet Coke but we'd run out of the crew diet drink on board. I got the cabin crew to give me one from the bar. A day or two later, I was surprised to get an email announcing: 'Scott, you can't really do that sort of thing!' Those cans of Coke cost the company almost nothing. All staff used to get a free drink on the aircraft when off on their holiday. I just needed to show my ID and the crew would not charge. These days, everyone is watching each other, making sure they don't cross a line. My view is: tell cabin managers what the onboard drinks cost and let them make the decision whether or not to offer them free.

In the old days in Britannia, all the crew bags would be taken to the hotel, I would tip the driver and our luggage would be taken to the rooms. The company expected us to tip the drivers who took us from the airport to the hotel, to build a good relationship so they would look after us if there were last-minute delays. By stopping everyone tipping, they saved between 50,000 and 70,000 dollars a year. But we had captains feeling embarrassed about not tipping, or tipping for the whole crew with their own money. The outcome was people lugging their own heavy bags all over the place. It was disrespectful and it was nonsense.

One of our aircraft – a B767 - was in Ibiza at a time when another flight found themselves a crew member short in Palma. Crewing asked the captain in Ibiza to fly the jet to Palma to drop off the required crew member. In the past, the Ibiza captain would have asked his cabin manager to bring him a volunteer, dipped into his ship's funds and asked our handling agent to take the crew member across to the terminal and get them on a scheduled service to Palma. But they didn't do that. In fact, when the captain pointed out that there was a scheduled Iberia flight going to Palma, he also had to tell our airline that he had no way of paying for the ticket. So the return flight to the UK had to go via Palma, at great cost. When the bosses sit down and look at these decisions, they look good on paper, but, at the end of the day, you have poor decisions dictated purely by money – and usually by incredibly small sums of money.

As captain, I always tried to look after my crew and that started by introducing myself and the first officer when I arrived at the jet. When integrated crew training (ITC) or joint CRM was introduced into the airline, there was a big push to improve the pre-flight briefings between the pilots and the cabin crew. Typically, the company would allocate no extra time for this briefing and crews again felt this was more lip service from the management. Comments about this came up in most of the classroom sessions I took when introducing the importance of the pre-flight brief. All I could say was it was important two-way communication at the beginning of the day and we should all try to achieve it. Of course, extra report time cost money in crew allowances and reduced the available flight duty time.

I used to tell them the flight times, any weather problems and if there might be turbulence. I told them just to call us if they wanted the 'fasten seat belt' signs on. I also suggested they give me a call if the temperature was not right in the cabin. My approach was: 'We'll all look after each other and we'll have a nice day out.' When I got off, even with the last passengers on board, I would pick up the PA mike and say: 'Thanks everybody for looking after us. Good job. Safe journey home.' The crew often told me they really appreciated that kind of thing. I knew they would be on my side and support me in my decision-making. Although management were often too thick to see it, it actually saved money, because when the crew said they were doing it for Scott, they knew I really appreciated their extra effort. I have no idea how you reverse this kind of attitude. I guess it would take a new CEO with charisma and an ability to convince staff this was the beginning of a new era. Maybe it will take another person like Freddie Laker to shake up the industry and upgrade the slack standards which have begun to pervade the levels of service. Someone who openly preaches about how their crews are treated and the knock-on benefits that has for passengers, the levels of service and the profitability of the airline. Perhaps that new airline boss will even say that anyone can phone them at any time of the day or night.

I am inspired by the idea of the very best trainers introducing and developing world-class training to aircrew. At JetBlue, it might have been their launch mentality for brand new people, but why not reintroduce that mentality and that approach every year? You've got to be able to say: 'Last year, this is what you asked for and this is what we've done. Here are some more ideas for you to think about. Would you all like your birthday off?' The crew have days off anyway so couldn't we just make sure one of them is on their birthday? That would leave them open-mouthed. If I were then to say 'What about if we were not going to fly Christmas Day this year, to give everyone time off with your families?', they would all think I was completely bloody barking. Management would quickly slag me off as a lunatic, a dreamer. I'd think to myself: 'Good. I'm glad about that. Because that's what I am.'

Starting the descent

Throughout my career, my life was flying the jet, and in the last twenty years I was also away instructing and examining in the simulator. One day that would all stop. One of our first officers once told me that he flew with a captain at a time when there had been voluntary redundancies on offer. This captain could have left the business with as much money as he was going to earn if he continued to fly until his retirement date, but he turned it down. Of course, this puzzled the first officer, who asked why he was flying this aeroplane, and he said: 'Yes, I could have left but then I wouldn't be a captain.' Which I thought was sad. I wondered if he wore the uniform around the house and if his pyjamas had four gold bars on them.

I was a captain and I'm proud of that, just as other people are proud about what they've achieved, but when I stopped being an airline pilot, I didn't lose all of that. I look back and think what a marvellous career I've had. First, I went part-time, primarily for Ruth and our relationship. I heard a radio programme about a chap who had no long-term memory and therefore lived in the now. It was clearly established that most stress is caused by thinking about past events. The other cause is thinking about the future and concerns about stressful events which are coming your way. At that time, I asked myself: 'What will cause stress in the future?' When I realised that the answer to that was the training department, I gave them six months' notice and quit. As a result, the emails stopped and all the things I didn't like about the department were gone.

I saw that footage of the aeroplane making an emergency landing on the Hudson River in New York after both engines failed due to multiple bird strikes. The captain was older than me, although he had a young family. I looked at that marvellous flying, thinking: 'Could I have done that?' People marvelled that he could pull off something like that 'at his age', but if you are qualified to command a jet, age shouldn't really matter - it is ability that counts. With age comes experience and that usually is a positive thing. Realistically, though, it does matter because I've seen, while examining, that age can make a difference in a very small number of pilots' ability to fly and operate in the simulator.

We all know elderly people who are bright as a button and others who are in mental decline. Physical decline is controlled by pilots' six-monthly medicals but mental decline will only show itself by mistakes on the line or on simulator checks. I have seen pilots over the years who passed their checks but, as they approached retirement, were marginal or indeed not acceptable. Adding five years to the flying life of a pilot, as ICAO did in 2006 by increasing the retirement age from sixty to sixty-five, was in percentage terms a lot. The change from sixty, as it had been for as long as I could remember, was quite sudden. The way we work and operate is busier than ever now. At sixty-five, many people are struggling with ailments and their hearing or eyesight might not be brilliant. There is no getting away from the fact that age has something to do with it, although the bottom line is when you come out of the simulator you are expected to cut the mustard.

Age does give you experience. For example, I would say to a new captain in training: 'After take-off, we are going to have a double engine failure - what are you going to do?' He might reply whatever the book says, instead of saying, as an experienced pilot would: 'I will turn left and crash on the beach – it's better than crashing a mile out to sea.' I've briefed my first officer, saying: 'We're on the grass on the next approach.' They think we won't need that and we'll get in on the next approach. But we might not. In the simulator, I would ask them what happens if ATC allows another aircraft to depart while you are out at 10 miles on your approach, short of fuel, and the departing aircraft has a brake or hydraulic failure and they need to be pulled off the runway. Do you have a plan B or even C? A plan B might be to go around, make a quick visual or radar circuit, then land. Plan C would be that if they hadn't managed to clear the runway, they would be landing on the grass. I would tell them: 'It won't be pleasant, but you will live.' There has to be a considered back-up plan. What about telling ATC to sterilise the runway?

I have heard first officers in the simulator briefing the captain about diverting to an alternate airport without realising they don't have enough fuel to make it with the gear and flaps stuck out. Their plan would lead to everyone dying. They say things which cannot have a good outcome. They've not thought it through, and that is what years of experience gives you. The Hudson River pilot knew he could reach the river but none of the nearby airports. He went from a normal day out to plan C in seconds. Most pilots wouldn't have got to the mental point of realising that a good outcome involved crashing into a river. That took experience and extremely good judgment. Many times, of course, the outcome after a bird strike like that would be even worse. If the aeroplane had been lower when the birds went into the engines, he might have said: 'I'm going to have to go for a park or between the buildings. There's going to be death and destruction but I am going to try to make it the least death and destruction possible.' As it turned out, because of his experience, his outcome was good.

I have tried to replicate that river landing in the simulator and I've let other pilots try it after a briefing and they inevitably hit the water hard and crash. As you touch down (ditch), you need to be in a very nose up attitude of 12 degrees because you are trying to get the tail of the aircraft to touch the water first. But with no engine power, that all has to happen just above the water and you have to get it all right first go, otherwise the aircraft will stall and plummet. So to give you a chance, you've got to come in quite fast. It doesn't matter where you splash down. You do have infinite distance on your side so you just let it nibble, nibble, nibble, nose up, up, up, up and splash down at the correct speed and attitude. Pilots I would expect to be able to do that made a bit of a mess of it. You can't really criticise them for it because it's hardly on the pilot syllabus – 'Execute a Hudson landing!'

So with age, sometimes you are exchanging steely-eyed, good aeroplane handling with experience. I have never been a fabulous handler of an aeroplane. There are pilots I have seen who can handle the aeroplane with more skill than I can, but I commanded an aeroplane pretty well. Most of the testing and development of new aircraft is done by ex-military test pilots. I have done some air tests with very good people but I never wished to go down that road because it's quite a numbers game. You're looking at flying the aeroplane with tremendous accuracy. I liked the idea of the airline pilot getting off the jet somewhere nice at the other end.

Some people would hate to examine in the simulator, whereas I got a huge amount out of that. I watched some examiners who could only say: 'Well, you were five knots fast and you didn't do that and you should have said that', which is no way to help a pilot. I've always felt pleased that I can see past that sterile approach and understand what is going through this chap's mind. How is he dealing with the co-pilot? How is he managing his and the other pilot's workload? How is his situational awareness? It is more important for the examiner to use judgment than wanting to see if the pilot busts arbitrary limits. Are they safe to fly our jets? It is more of a mind game than many people realise. Currently, that way, the more mature way in my view, is being devalued. It is going back to numbers.

The company introduced a requirement to score the pilots in the simulator using a system on a laptop, which I think is not the best solution. I would want the examiners to sit back and watch and make notes. They also have to operate the simulator, be air traffic control, be cabin crew engineering and anyone else the crew wish to communicate with. This new system is all about gathering data, and data is based on numbers. I know the good guys ignore most of this and complete the computer information back in the hotel, but less capable trainers will sacrifice judgment for data gathering. One of the problems with working for a training organisation or simulator provider, such as Boeing, is that the training and checking you provide is purely training by numbers. You don't know the airline's culture and their standard operating procedures (SOPs), so the outcome is that you are remote from the students and you are giving them very basic training. I did that, years ago, when I was new and trying to get on top of the simulator panel.

I was getting used to teaching in the simulator but I did not find it very rewarding. Once, I trained Turkmenistan pilots through an interpreter - you can imagine what fun that was! Boeing asked me to come and work for them straightaway on a part-time basis, and then to join them full-time when I retired, but I was not interested. Training within an airline when you know the culture of that airline and the SOPs, enables you to give good feedback and hopefully some useful tips. If I wanted to do something else, I would be thinking more along the lines of setting up a world-class training management team, which could go anywhere in the world and help airlines and corporate jet operators that might be struggling with their training. They might also be struggling with the introduction of a new aircraft type or need to train and develop their own training pilots. I am confident there would be a call for this expertise. I could put together a team of world-class trainers and managers with airline jet and national civil authority backgrounds.

On the other hand, I was asked if I was interested in the job of chief pilot of a B747-400 cargo operation in Germany a few years ago. I thought: 'Why would I want to do that? I want to take my foot off the accelerator.' I would have just been going to deal with the kind of stuff I had dealt with before. There must have been problems if they wanted a Scotsman to come and sort out their fellow Germans! It might have been fine for a couple of years but then I would certainly get fed-up with it.

All of my career I did nothing but move up a ladder. Now that ladder is gone. I've kicked it away. One thing is quite certain: I will continue to fly light aircraft as long as I can. I actually thought until recently that I might renew my flying instructor's rating. On second thoughts, it became clear that I would be dealing with all the rules, regulations, courses and tests, and dealing with the Civil Aviation Authority. I still have my share in a light aircraft and even if I sold that, I would continue to fly out of Turweston Aerodrome. That sounds pretty good to me.

Maturing with age

I've been lucky to have found a group of close friends who are in the aviation business. You are unlikely to find me in the crowd at the bar, although it was different when I was in training management. I would be very comfortable in the middle of a bunch of my trainers. I have always enjoyed the camaraderie, the teamwork of a crew. Sometimes you get good crews, sometimes you don't get a crew you particularly enjoy socialising with. It doesn't matter if you are best buddies with these people, it is always important to engage with the crew, especially the big players such as the cabin manager and the first officer. Back in the old Britannia days, it was the 'boys and girls' because there were no stewards. Only 'boys' were the pilots, though we did have one woman first officer when I joined in 1982. Of course, our lifestyle back then was that we would go away for five or six days and stay as a crew. We ate as a crew and went out as a crew. That was the norm.

It was a completely different world to the one my brother knows, in an airline like British Airways. They are, in every sense of the word, an old and established airline. When I was a young pilot, most of the crews were young and it was fun and exciting. At British Airways, the cabin crew have been there a long time so their average age is bound to be quite a bit older. These days, if pilots go away with a Gatwick crew who do a lot of long haul, they don't all tend to go out at night. Some of the crew will be older, they've done it and seen it countless times. A Newcastle crew, on the other hand, who don't do a lot of long haul, will see things very differently. If you take them to Orlando, they will want to go and play - and that usually includes wall-to-wall shopping as well. But I really liked the camaraderie of those trips. I have flown to the Far

East and Australia, and been away for a few weeks. It was just so much fun. When I did my command, the whole accent was on not hitting the hill. The crew were almost taken for granted, but it has always been my style to enjoy leading the crew and supporting them. Not all pilots see it that way. I have been in far-off places when a member of another cabin crew has come up to me and said: 'One of our crew members is in hospital and our captain has disappeared off for the day with his wife.' That lack of responsibility to the crew is unacceptable. I would feel, like most captains, that my place would be at the hospital and then to inform the crew, and the company if necessary. I felt I had that responsibility and, as a consequence, if the crew let me down, maybe I took it harder than others. During my time as a first officer, before any of us had heard of CRM, I saw captains who just did not know how to relate to crews, who were often younger than their grown-up children. On arrival at the hotel, the cabin manager would sometimes say to our miserable captain and me that they were having an early night. When I got to my room, there would be a phone call to tell me where they were meeting for drinks.

I soon found that when I had to get cross with a crew, which was very seldom, it probably indicated that I had run out of ideas. In a way, I felt I had failed. The idea behind integrated crew training was to learn about and understand each other's roles and expectations. With the pilots and the cabin crew in the classroom, we would be talking about emergency procedures and what was happening on the other side of the flight deck door. There was also the question of the authority of the captain and when to use it.

For example, it would not be appropriate for the captain to say to the crew: 'I insist that we all go off for a Chinese tonight.' But if we were in a Muslim country, it would be correct for the captain to have a chat with the cabin manager about dress codes and ask them to have a quiet word. In this case, it's less about the authority of the captain and more about his role as manager of the crew. CRM was never just about crew getting on well with each other. It teaches the skills required when the crew are not getting on well and supporting each other. These are the skills to cut through personality problems, which can be bad for flight safety. It is about giving people tools to deal with problems and advise them when it is best to say something and when it is better to say nothing. It was an attempt to standardise how people managed similar situations. The captain had to be the boss but didn't want to find himself in a situation where he had to start saying: 'I am the captain and I am telling you...' There is nowhere to go after that. Fortunately, I was never caught out like that, although there were a couple of times when I came close. I do know captains who have not enjoyed the crew handling part. Frankly, I think they would rather have flown freight for a living. They found it a struggle and I think that was down to their personality. They were good pilots but somehow they missed the point that they were working for a commercial airline. They forgot that our passengers needed a motivated and happy crew. I regarded the cabin crew as my link to the passengers. That was why, when they were having a bad time in the cabin, I'd try to deal with the passengers directly and speak to them. This is a little more difficult than people might imagine these days because of mobile phones and the ability to film and record. If they were recording what I was saying on their phones, saying 'Turn your phones off' wasn't going to help the situation. I'd go and speak to the passengers because I felt I had a responsibility to them and I felt comfortable doing it. It was a positive thing if I could get them on my side. When I walked through the aeroplane, I would have a little chat and a joke with passengers - and I usually even got a little round of applause. I knew I was looking after my crew because the passengers were not now going to pick on them.

My role on the aeroplane, when flying an airliner, was always going to be different to when I go flying in a little airplane by myself. As soon as someone is with me in a little light aircraft, I have another responsibility. On one occasion, I made someone sick doing spinning in a light aircraft over Loch Lomond. After recovering from the first spin and climbing back to altitude, I said: 'Was that okay?' He just smiled and said nothing. I asked if he wanted to do it again, and again he smiled. It turned out that his smile was actually a fixed grin - because he was so scared. He definitely didn't want to do it again and he threw up all over the front of the aircraft. I was the one who was insensitive to the comfort of my passenger. I would never push so hard that I would scare someone in a car. In a car, they can say: 'Stop. I'm getting out.' You can't do that in an aeroplane.

I have understood from the beginning of my airliner career that I have a responsibility to every person in the aeroplane. When I was flying the jet, it was taken as read that I didn't want to hit the hill and the first officer didn't either. But the crew sitting in the back had absolutely no control over the situation. They were completely relying on us at the front to make whatever was going to happen as survivable as possible. That was my job - to make the outcome as good as I possibly could.

I have loved the camaraderie of the whole experience. It was the same during my early years with Loganair, even though it was either just me and the passengers, or the captain, me and the one cabin crew. Later, climbing aboard a B767 with 11 people in my crew was something I always enjoyed and it was disappointing if it didn't go well. That didn't happen very often.

I've also loved the sheer awesomeness of looking out of the flight deck window on the Earth beneath us. It is pretty obvious but I have seen some amazing sights, just as all commercial airline pilots must have. I will never forget the moon rising over the Atlantic. Because we would be at height, the moon was below us because of the curvature of the earth, so you would be looking down at the moon rise. It is most spectacular when coming back at night, heading east, when the moon comes up in our two o'clock position. I'd suddenly see a wee, faint glow and then I'd watch the moon come up quite quickly from our dark cockpit. We'd usually darken the cockpit lights so we could do a bit of stargazing as well. Before the moon rises, the sky will be quite dark and if I tilted my head over at my side window, I could look straight up at the Milky Way, that faint almost mist-like grouping of billions of stars. That was something I never got tired of looking at. First officers often had a star chart with them to identify what is what.

Another awe-inspiring sight which never fails to impress is looking down across Greenland at the ice floes, the glaciers and the pack ice. The first time I saw an iceberg was crossing the Atlantic decades ago with Jock Adams as we flew to Goose. I was only a young pilot and I asked: 'What's that in the water?' As we got closer, I realised it was a huge iceberg. I have crossed the Equator quite a few times, and every time I found the southern sky breathtaking. There is something amazing about the view as we crossed India on the way to the Far East. These sights are what I missed when I stopped flying.

Closer to home, coming into Luton might not sound the most glamorous approach in the world, but because it is just north of London, coming in on a clear night over the city is spectacular. Deep in the night, there is not much traffic about and nothing going into Heathrow so ATC would drop us quite low over London. Even coming into Birmingham is interesting because we would come over Ostende or Rotterdam and cross the Thames estuary, heading for Cambridge and then across to Milton Keynes, flying directly over my home town of Olney. Sometimes, at the start of the descent over the Channel, you can see right up towards Newcastle on a clear night and at other times you might even be able to see Dublin. It is unbelievable sometimes because, at height, you can see for 200 miles and, at night, light really travels. One can never get enough of that. I never got bored with those experiences and neither do the other pilots I have spoken to. Passengers love it too, of course. If I said 'Look out of the left-hand window…', they would all be craning their necks, particularly if they had not seen a particular view before. Crossing the Alps, you will usually look down and see the peaks covered in snow and the odd glacier, and occasionally spot the Matterhorn.

It is beneficial to be reminded of the magnificence of the planet. I never ceased to be delighted as we came over the Empty Quarter, having crossed the Indian Ocean on the way to Jeddah to the west. When you hit this Empty Quarter, it is just desert. It looks like a vast beach after the waves have washed over it. The sands are undulating but there is absolutely nothing there. Heading into places like Denver in the mid-west of the US is fascinating, as is flying over New York and other major cities at night. Over Paris in the day, you can look down and see the Seine and the Eiffel tower. Of course, not every location looks pretty from the air. No matter how many times you see places like Lanzarote or Reus, they never look any prettier. Lanzarote, to me, always looks like the moon – it's just dust. There is nothing memorable about that place. What a contrast to the view of the Taj Mahal from 36,000 feet. Just the fact that you can find it and see it from such a height is always so thrilling.

I have sometimes been asked what perspective all this gives me on the planet as a whole. The simple answer is - the vastness of it. When I crossed the southern Pacific Ocean on the round-the-world tour, it took me three days to cross it in a jet. Not a leisurely three or four hours… three days. When you look at the globe and you spin it around, you will notice that the South Pacific extends from the right-hand edge to the left-hand edge of the globe. It is colossal. This planet is so incredibly big, and when you are up there, you can't help but reflect how puny mankind is compared to this hugeness. Our place is put into real perspective when you look up at all the stars which, most of the time, we can't see on the ground because of the light pollution from the towns and cities. As we pass over London, the lights look magnificent but looking up to the Highlands of Scotland, it's completely dark. We fly over countries such as Portugal and most of it is just black, but over the south of England there are loads of lights. It is all pretty awe-inspiring when viewed from the flight deck. From the passengers' point of view, it is not the same. Passenger windows are not very clear and quite small so they don't get the impression of hugeness that we do in the flight deck, where the windows are bigger and the views are more panoramic. I never looked at the North Star without identifying the

Plough, and for no particular reason took the opportunity to check my compass with the star. It wasn't that many years ago that the flight deck used to navigate by the stars. The navigator would use a sextant to navigate by before inertial navigation and GPS. In my time with the Jumbo, I was really pleased that I had the chance to fly with a flight engineer. I was always fascinated by the idea of a third person on the flight deck, who was not a pilot but still an integral part of the flight crew. I can only imagine what it was like years ago with a flight engineer, a navigator and a radio operator. The flight deck must have been more like the bridge of a ship.

There are only a few aeroplanes which have really made a lasting impression on me, foremost of which is the B767, a truly exceptional plane. When I first saw the B747 Jumbo, I was impressed by its sheer size. The thought that I was going to fly that huge aircraft did bring some trepidation at first but, as a pilot, you have to just get on with the business of flying. When you do, the size of the thing is not an issue. In fact, the monster turned out to be pretty under-powered. If I mention that I've flown the Jumbo, many pilots have said: 'How lucky was that?' I've had to reply, quite realistically: 'Actually, it's just a gutless B767.' At the same time, I'll tell them: 'But you've got to experience it for yourself, if you can.'

The first time I flew the Jumbo, we all arrived at the aeroplane in a coach which stopped at the bottom of the steps because it wasn't parked on a jetty. It was awe-inspiring. So much so, that I deliberately averted my gaze, scampered up the steps and then, without looking left or right, up the spiral steps and into my seat. Was I nervous? You bet. Was I surprised? Yes, because it turned out that it flew just like any other aeroplane. But that experience, from a pilot's point of view, certainly does tick some boxes.

It's not all about speed. I always enjoyed taxiing, especially in the big jet. You couldn't actually see your wing tips from the flight deck so I sometimes only had a vague idea of where everything was. I knew where the left wing tip was by looking through my side window, past some screws set in the frame and I could line it up pretty accurately. If I ever thought my wing tip was getting close to something, that was the time to stop and request a marshaller. Having a good idea of where your outer wheels are is important if you need to turn 180 degrees on the runway. In fact, you can't turn a B767-300 on the runway because the standard runway is not wide enough. You can turn a B757 around but you will need to know where the outboard wheel is, to make the most of the turn without hitting any lights down the edge of the runway.

The first sectors of command training are the first opportunity for new captains to taxi the aircraft, as there is no tiller on the right-hand side, so first officers can't do it. If they damaged their aircraft, there would be all hell to pay. If they damaged it while I was training them, I would be getting the bollocking.

Taxiing has not really changed in the last 40 years. There are a few airports with ground radar, which is really more for foggy conditions. Some major airports have fancy taxi lights for low visibility. At the larger airports, including Birmingham, pilots can simply follow the lit green lights that lead them to their gate, but they are still few and far between. Some airports only have runway edge lighting, so at night as you flare the aircraft, you are looking out on a dark world, lit only by the beam of your landing lights. Modern aviation still has its challenges, in the air and on the ground. There's rarely a dull moment.

When I went into the business, I knew airline pilots earned a reasonable salary and I have been very lucky in that regard. I've never wanted financially. I'm not really one for parties, but people always seemed keen to ask me about flying, which made things pretty easy for me. When someone said 'Oh, you're an airline pilot', they immediately assumed I was more intelligent than I am. I've had a pretty easy ride of it, socially and economically. Once, taxiing in at Birmingham in the B767 one very dark, winter night, it was cold and wintry outside and raining. I was on the flight deck, all warm and cosy in a short-sleeve shirt, and I saw some workmen huddling in a hole in the ground next to the taxiway. They were wearing oilskins, working in really miserable conditions in the dark in a hole in the remotest corner of an airfield. I suddenly thought to myself: 'How lucky am I? This has got to be better than the lot of most guys.' Some people say: 'I'd hate to be in your shoes, living that kind of life and carrying such responsibility.' None of that fazed me one bit. I can look back on my career and contemplate how the business of flying aeroplanes has given me so much pleasure. I am seen as successful, lucky and fortunate, and it has been fabulous, quite fabulous.

The end

My last flight on the jet was in August 2015. A few days before this flight, I was pilot monitoring on a sector into Kos at night. The airfield is a dark hole, on a dark island - not an easy approach. My first officer was doing a good job, but I was struggling to focus on my let-down charts and monitor carefully the descent profile of our height vs distance to touchdown. A few days later, I landed into Larnaca, again at night, and as we cleared the runway, ATC gave us a string of letters indicating the taxiways they wanted us to use to get to our stand. I just couldn't read the taxi chart. The old paper charts had recently been removed and replaced by a computer screen on my left side. I couldn't just lift the paper chart up to my face and read it, as you see people doing when they have very bad eyesight. My first officer read back the taxi instructions and guided me in. I felt very uncomfortable relying on my colleague without me checking as well. My corrected long sight was OK, but I had been struggling with my mid-distance, the distance to the flight instruments, especially the flight director. I had always been okay with reading the paper charts because I could unclip them and read them up close.

I was pilot flying on the return sector out of Larnaca. The take-off was fine until I rotated and my scan then dropped down on to the flight director. It was very blurred, and the only way I could focus was to close my right eye. As I climbed away, I thought 'this is not good'. The next day, I made an appointment with my AME (aviation medical examiner) and took myself off my next flights. The AME grounded me, so I went to the optician for an eye test and then my GP, who referred me on for further tests on my binocular vision. I was moved on to PHI (permanent health insurance) and received LoL (loss of licence

With some of the crew at Zakynthos before my final flight

insurance). I am a lucky boy. I was 61 and had flown professionally for 37 years and privately for 46 years. I could hang up my flying gloves if I had to. Quite neat, really: the book ends as I end my flying career.
I don't miss getting up in the wee small hours, my checks in the simulator, the annual ground exams and line checks, landings at 6am with the sun burning my eyes, coasting out over Canada at 1am with the Atlantic in front of me and a landing at breakfast time, the long drive back from Birmingham after a tiring flight. I DO miss bursting out of cloud into a bright blue cloudless sky, looking up at the Milky Way mid-Atlantic, and the exhilaration of the take-off run. But casting off the shackles of a roster, spending time with Ruth and the family, trips in our campervan, getting the old saxophone out again… that makes me smile.
As I have said before, I feel truly blessed, and long may it continue.
Now I'm retired, what have I left behind? An industry that has never been safer, with technology continuously forging ahead, a worldwide terrain database for GPS accuracy down to a few feet, satellite communication and automated flight – the list goes on. But what of traditional pilot skills? These are still vitally important for today's airline pilot. We need highly trained, highly skilled pilots flying our aircraft in each and every cockpit. I believe our industry will continue to strive to develop new and better ways of training, motivating and empowering these men and women to keep us all safe. After all, it is their drive, effort and skill that will continue to ensure that the skies above our heads remain a safe and joyful place for us all to inhabit.

Printed in Great Britain
by Amazon

68940584R00208